有机配体保护的钯纳米粒子的制备及其分析应用

◎ 张磊 著

中国农业科学技术出版社

图书在版编目（CIP）数据

有机配体保护的钯纳米粒子的制备及其分析应用 / 张磊著. -- 北京：中国农业科学技术出版社，2025.1.
ISBN 978-7-5116-7091-5

Ⅰ.TB383

中国国家版本馆CIP数据核字第2024FQ1233号

责任编辑	王惟萍
责任校对	王 彦
责任印制	姜义伟　王思文

出 版 者	中国农业科学技术出版社
	北京市中关村南大街12号　　邮编：100081
电　　话	（010）82106643（编辑室）　（010）82106624（发行部）
	（010）82109709（读者服务部）
网　　址	https://castp.caas.cn
经 销 者	各地新华书店
印 刷 者	北京捷迅佳彩印刷有限公司
开　　本	170 mm×240 mm　1/16
印　　张	8
字　　数	145千字
版　　次	2025年1月第1版　2025年1月第1次印刷
定　　价	46.80元

版权所有·侵权必究

前 言
FOREWORD

纳米材料被人们誉为"21世纪最有前途的材料",从而受到科研工作者越来越多的关注,因其本身所具有量子尺寸效应、宏观量子隧道效应等特殊性质,展现出的许多优异的特性,使其具有广阔的应用前景。钯作为重要的铂族元素和过渡族金属之一,在催化、光吸收和磁性能等方面体现出特有的物理和化学性质,而这些性质与其的微观结构有着千丝万缕的联系,制备稳定性、水溶性和分散性好的钯纳米材料已经成为研究的发展趋势,因此,针对钯纳米材料性能、结构分析方面的研究具有重要的意义。

本书合成了以有机配体N-乙酰基-L-半胱氨酸(NAC)和N,N-二甲基甲酰胺(DMF)为配体保护的小尺寸水溶性钯纳米粒子;通过紫外-可见吸收光谱、红外光谱、透射电子显微镜、X-射线光电子能谱、热重分析、动态光散射等技术对钯纳米粒子(Pd NPs)进行物理和化学性质的表征;并结合反向-高效液相色谱法对所制备的钯纳米粒子进行了分离分析研究,利用质谱等手段得到钯纳米粒子更加丰富和详细的化学组成信息;此外,将水溶性钯纳米粒子修饰电极应用于检测重金属铜离子,获得具有灵敏度高、选择性好、稳定性优良、成本低廉的电化学传感器。

本书研究的创新之处,第一,建立了一种有机配体稳定钯粒子的制备方法,该方法具有一定的普适性,可以推广应用到其他有机配体或者其他贵金属纳米材料的制备过程中;第二,首次将高效液相色谱法应用于钯纳米粒子组分的分离,并利用紫外-可见光谱、荧光光谱和质谱法测定钯纳米粒子组分中所包含的钯原子和配体数目,建立了一种分析钯纳米粒子组成的有效方法。

但是,钯纳米粒子的应用与环境分析化学方面仍然是一个有待深入的课题,虽然本书使用不同的有机配体所做的尝试，为实际运用提供了积极的帮助。但是，在将来的研究工作中还需继续努力，特别是对DMF-Pd NPs/GC修饰电极的检测机理进行更进一步的研究。

最后，我们感谢所有为本书提供帮助和支持的人。同时，我们也希望本研究能够为未来的相关研究提供参考和启示，共同推动钯纳米材料研究领域的不断进步。

著 者

2024年8月

目 录
CONTENTS

第1章　绪论 ·· 1

　　1.1　纳米材料的特性 ·· 3
　　1.2　钯纳米粒子（Pd NPs）的制备方法 ·· 5
　　1.3　配体对Pd NPs的影响 ·· 7
　　1.4　Pd NPs的表征手段 ··· 12
　　1.5　Pd NPs的应用研究进展 ·· 17
　　1.6　立题背景和研究内容 ··· 19
　　参考文献 ·· 22

第2章　NAC保护的Pd NPs的制备和表征 ·································· 35

　　2.1　引言 ·· 37
　　2.2　实验部分 ··· 38
　　2.3　结果与讨论 ··· 40
　　2.4　结论 ·· 45
　　参考文献 ·· 46

第3章　NAC保护的Pd NPs的色谱分离分析研究 ······················· 51

　　3.1　引言 ·· 53
　　3.2　实验部分 ··· 54
　　3.3　结果与讨论 ··· 56

3.4 结论 ········· 64
参考文献········· 64

第4章 DMF保护的Pd NPs的制备和表征 ········· 69

4.1 引言 ········· 71
4.2 实验部分 ········· 72
4.3 结果与讨论 ········· 73
4.4 结论 ········· 78
参考文献········· 79

第5章 DMF保护的Pd NPs的色谱分离分析研究 ········· 85

5.1 引言 ········· 87
5.2 实验部分 ········· 88
5.3 结果与讨论 ········· 90
5.4 结论 ········· 98
参考文献········· 98

第6章 DMF-Pd NPs修饰的玻碳电极对Cu^{2+}的电化学检测 ········· 103

6.1 引言 ········· 105
6.2 实验部分 ········· 106
6.3 结果与讨论 ········· 107
6.4 结论 ········· 111
参考文献········· 112

第7章 总结与展望 ········· 115

7.1 总结 ········· 117
7.2 展望 ········· 119

彩图 ········· 120

第1章

绪　论

纳米科技是在纳米尺寸范围内认识和改造自然，通过在原子和分子的层面操作来创造新物质，使材料被形成独特的规模、结构和组成的一门科学[1]。"纳米粒子"最初称为"小粒子"，是在20世纪80年代提出。"纳米"是肉眼不可见的微小尺寸，超出了我们观察的正常范围，一般认为，尺寸小于100 nm都可以称为"纳米材料"[2]。图1.1展示了纳米粒子长度尺寸的分类。纳米材料被视为原子和块体材料之间的桥梁，并且它们在光、磁、热、电等方面的性质发生很大变化[3]，具有普通材料不具备的奇异和反常的性质。迄今为止，科研工作者已成功制备各种类型的纳米材料，包括磁性纳米材料、碳纳米材料、贵金属纳米材料，其中贵金属纳米材料因其独特的物理化学性质，已然成为化学、物理学、生物学、医学和材料科学领域中的研究热点。

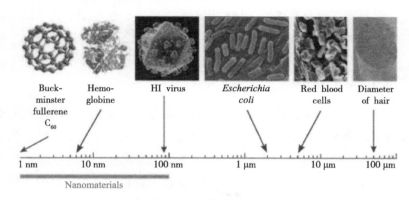

图1.1　纳米粒子长度尺寸的分类[1]

1.1　纳米材料的特性

　　纳米材料主要由纳米晶粒和纳米晶界组成。纳米材料最突出的结构特征是尺寸很小，而晶界原子的比例大。这些结构特征使纳米材料具有一些特殊的性质，如小尺寸效应、量子尺寸效应、表面界面效应和宏观量子隧道效应等[4]。

1.1.1　小尺寸效应

　　当纳米粒子的尺寸与物质波波长、光的波长以及超导态的相干波长或透射深度等相当或者更小时，粒子与波之间的相互作用不同于大尺度粒子，此时由于粒子晶体周期性的边界平衡条件遭到破坏，其表面层周围的原子密度变小，从而使材料的光、热磁、电和力学等性质发生变化，产生了一系列奇异的现象和性质[5]，这些由材料尺度变化而引起的物性变化被称为小尺寸效应。例如，材料的尺寸减为纳米尺寸时，吸收光的位置会发生移动，当纳米材料表面有吸附聚合物时，其共振峰会出现移动或峰形产生变化[6]，可用于判断纳米粒子是否与聚合物吸附。纳米材料的热学性质也有不同，对于块状贵金属来说，它们的熔点是固定的，小尺寸效应使材料的熔点降低，贵金属金的熔点是1 064 ℃，而10 nm的金纳米粒子（Au NPs）熔点为1 037 ℃，降低了27 ℃[7]；某些纳米材料的磁性与块状材料也有显著变化，在纳米尺寸条件下，纳米粒子呈现超顺磁性的性质[8]，可广泛应用于磁性钥匙、磁卡及电子器件中。在通常情况下，陶瓷材料具有耐高温和耐强腐蚀性的优点，但是陶瓷也具有脆性、韧性比较差等弱点，通过纳米技术制成的纳米陶瓷，韧性得到大大的提高，是解决陶瓷脆性的方法之一[9]。

1.1.2　量子尺寸效应

随着粒子尺寸的变化，粒子的电子结构也随之发生变化，金属附近的电子能级从准连续能带转变为分立的电子能级，当能级间距大于磁能、热能、光子能量时，从而导致纳米粒子的热、磁、光、力等性质与宏观物质特性有显著不同[10-12]，即为量子尺寸效应。如量子尺寸效应可使金属纳米材料拥有不导电的性能，还可以让二氧化硅纳米材料由绝缘体变为导体。当金属纳米粒子粒径小于光波长时，会失去贵金属原本的光泽而呈现出黑色，使金属纳米颗粒子降低对光的反射率，可利用此性质使用贵金属纳米材料将太阳能转化为热能和电能，也可以应用到红外敏感电子元件和军事中红外隐身技术等领域中。

1.1.3　表面界面效应

当材料粒径的减小时，会使纳米粒子表面的原子数与总原子数的比例增大，由此引起材料性质的变化，称其为表面界面效应[5]。材料的比表面积与粒径成反比，随着颗粒粒径的变小，其比表面积会显著增加。当纳米粒子达到纳米尺度，表面原子所占比例增加到50%以上，表面能也会迅速增大，使表面原子具有很高的活性，同时也极不稳定，造成纳米粒子具有很高的活性，可用于助燃剂、高效催化剂和电子器件等。例如，汽车尾气排出前，在钯纳米材料的催化作用下，可将有害的成分一氧化碳、氮氧化合物和碳氢化合物等转化为二氧化碳、氮气及水蒸气等，转化率可高达90%[13]，因此，钯纳米材料在汽车废气的处理中有举足轻重的作用。在表面界面效应的作用下，可将纳米粒子作为气相色谱柱的固定相[14]，从而可选择性的获得带电荷的目标物。

1.1.4 宏观量子隧道效应

隧道效应是微观体系中一个独特的现象，是指能量小于势垒高度的粒子有越过势垒的可能性。科研工作者通过研究纳米材料体系的宏观物理量，发现纳米粒子也具有磁化强度等隧道效应，因此，将之称为宏观量子隧道效应。如在低温下，超细镍纳米粒子仍可以保持超顺磁性。量子尺寸效应和宏观量子隧道效应在纳米粒子做基础研究以及工业中实际应用时，其重要性显而易见，例如，它们会影响仪器信息存储时间极限和现有微电子、光电子器件的微型化的程度[15]。

1.2 钯纳米粒子（Pd NPs）的制备方法

Pd NPs是重金属纳米粒子中稳定的纳米粒子之一，其具有良好的稳定性、小尺寸效应、表面效应和良好的生物相容性等特点，从而得到广泛的应用。Pd NPs作为催化材料以及其他一些应用，如在偶联反应、催化加氢和传感等方面具有非常重要的作用，它们的合成深受科研工作者的关注。Pd NPs作为稳定的贵金属纳米颗粒之一，制备方法的不同，所合成Pd NPs的形貌、粒径分布和晶型结构也会不一样，主要可分为物理法、化学法和生物法。

1.2.1 物理法

物理法（自上而下）是贵金属纳米材料最早采用的制备方法，其原理为采用高能消耗的方式，将固体金属钯经粉碎、研磨等物理方法强制细化得到纳米材料，或用人工的方式将钯原子组合为纳米粒子。常见的物理法有蒸发冷凝法[16-17]、高能球磨法[18]、离子溅射法[19-20]、等离子体沉积法[21]等。物理法制备钯纳米的方法优点是得到的钯纳米产品的纯度高、操作方式简单，但是对仪器设备条件要求高，在制备过程对仪器的损耗比较大、产量低，并对所制备的纳米粒子的形貌也无法掌控，因此，物理法并不是制备Pd NPs理想的方法。

1.2.2 化学法

化学法（自下而上）是指将钯离子采用热分解、化学还原、超声波辐射、微波辐射和电解等方法形成钯原子，通过不同数目钯原子的聚集、组合形成Pd NPs。其主要方法包括热分解法[22-25]、超声波辐射[26-28]、微波辐射[29-30]、光化学[31-33]和化学还原法[34-40]等。与物理制备法相比，化学法操作简便、成本低、合成过程可控性好、所制备的Pd NPs的形貌可控、粒径尺寸小和分散性较好。

1.2.3 生物法

生物法，又叫生物还原法，是使用生物质中具有还原性的基团将金属钯离子还原为钯原子，其中微生物、植物提取液等作为还原剂合成Pd NPs。

1.2.3.1 微生物还原法

微生物还原法是利用细菌、酵母菌中细胞含有的还原性基团，还原钯粒子，制备Pd NPs。雷彬等[41]采用克雷氏杆菌作为还原剂制备Pd NPs，经红外光谱（FTIR）、X-射线衍射（XRD）、透射电子显微镜（TEM）等表征，合成的Pd NPs为球形，其平均粒径约为15 nm，并将Pd NPs催化六价铬还原研究其催化剂性能，研究表明，克雷氏杆菌表面的氢酶对Pd NPs表面的活性具有调控作用，因此，极大地提高了催化剂的催化性能。与化学法相比，微生物法可以降低环境污染等优点，但是微生物培养、生长不可控、还原率低，需要进一步深入研究。

1.2.3.2 植物还原法

植物还原法是指在温和条件下，利用植物质来制备金属纳米颗粒的方法，其具有生物相容性高、降解性快等优点[42]。植物质中的还原糖、类黄酮、多元醇和氨基酸等生物化学成分都可参与贵金属离子被还原为零价金属和成核过程，从而成为不同大小和形状的金属纳米颗粒[43-44]。Dauthal等[45]利用龙葵叶提取物合成Pd NPs，并对其2-硝基苯酚及2-硝基苯胺的催化加氢活性进行了评价。通过该方法合成的Pd NPs的粒径在2～4 nm，并且表现出良好的催化活性。贺媛媛等[46]利用植物还原法，以二价钯离子为前躯体，银杏提取液为还原剂制备Pd NPs。电感耦合等离子体原子发射光谱法（ICP-ACE）测定溶液中的残余Pd离子浓度进行定量研究，结果表明，当Pd离子浓度越大，银杏叶颗粒越小，其还原效果越好。秦聪丽等[47-48]采用桑叶和柚子皮提取液作为还原剂制备Pd NPs，通过FTIR、X射线光电子光谱（XPS）、TEM、扫描电子显微镜（SEM）对合成的Pd NPs进行表征和分析，研究表明，所制备的Pd NPs粒径均一、分散性较好。

生物还原法制备Pd NPs具有原材料丰富、反应条件温和、污染低等优点，成为近年来制备纳米颗粒的热点，但是不同生物质中的还原性官能团的作用机制不明确，其制备的技术还处于实验室研究阶段，需要科研学者进一步地探索和研究。

1.3 配体对Pd NPs的影响

在制备过程中，为了防止Pd NPs发生团簇和聚集，得到形貌良好、分散性和粒径尺寸均一的Pd NPs，人们一般在合成过程中加入配体作为稳定剂，一般最常见的稳定剂有有机配体、表面活性剂、聚合物等（图1.2）。所合成Pd NPs的性质、性能与其所选用的配体息息相关，按所选用配体的不同，对Pd NPs的影响总结如下。

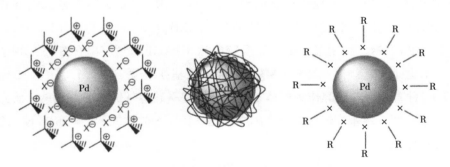

图1.2 使用不同的配体合成Pd NPs[49]

1.3.1 巯基化合物

铂族金属与软性硫基供体之间的强相互作用使含硫配体的巯基化合物成为纳米粒子的高效稳定剂。Brust等[50]在1995年用四辛基溴化铵（TOAB）作为相转移试剂,把金离子转移到甲苯有机相中,通过十二烷基硫醇对金纳米颗粒进行保护,使用硼氢化钠(NaBH$_4$)还原金离子得到金纳米粒子,用此合成方法合成的金纳米粒子对热和空气稳定、粒径可控,并且所制备的纳米粒子如同简单的化合物,可复溶,可干燥且不影响其性质,证明硫醇在金纳米粒子的两相制备中是一种优异的稳定剂,该方法已用于制备含有各种贵金属(包括钯、铂、银等)的纳米粒子,且粒径1~3 nm的重金属纳米颗粒通过硫醇盐配体的单层稳定,在普通有机试剂中不凝聚、不分解,能够像有机化合物一样易被处理和官能集团化。

烷基硫醇保护的Pd NPs可以通过Brust两相方法直接制备,图1.3[51]为使用改进的Brust方法合成硫醇稳定的Pd NPs的合成示意图,其中S-R为不同的烷基硫醇配体。四氯钯酸盐通过加入长链铵盐(例如,四正辛基溴化铵或季铵氯化物)将其从水相转移到有机溶剂(如二氯甲烷或甲苯)中,再加入配体后,用NaBH$_4$水溶液还原有机相中的钯离子,从而合成配体保护的Pd NPs。另外,长链硫醇分子通常作为配体,Sadeghmoghaddam等[52]用S-十二烷基硫代硫酸盐作为保护Pd NPs的配体,合成Pd NPs,也被证明是一种有效的替代方法。Zamborini等[53]利用Brust两相法制备十二硫醇和己硫醇为配体保护的Pd NPs,通过核磁共振(NMR)、FTIR、紫外-可见光谱(UV-vis)、TEM和热重分析(TGA)等表征技术对合成过程和Pd NPs进行分析,研究表明,可以采用改变原料中硫醇和钯离子的摩尔配比来控制产物的粒径大小等性质。Sakai等[54]也采用谷胱甘肽(GSH)为配体,NaBH$_4$作为还原剂,当[GSH]/[Pd^{2+}]摩尔比小于3:4时,所合成的为Pd NPs,利用聚丙烯酰胺凝胶电泳(PAGE)测定出含有15~39个钯原子。使用11-巯基十一烷酸(MUA)、9-巯基-1-壬醇(MN)、正十二硫醇(DT)3种直链烷基硫醇作为配体,合成不同烷基硫醇基保护的Pd NPs,研究表明,反应时间、巯基配体与钯离子的摩尔比、还原剂的添加量影响着Pd NPs的平均粒径和分散性。

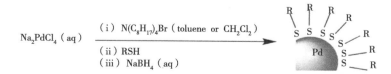

图1.3 使用改进的Brust方法合成硫醇稳定的Pd NPs[51]

1.3.2 表面活性剂

表面活性剂作为重金属纳米粒子的稳定剂在合成过程中也被科研工作者所关注。凡加入少量而能显著降低液体表面张力的物质，统称为表面活性剂。按极性基团的解离性质分类如下：阴离子表面活性剂十二烷基磺酸钠（SDS）、十二烷基苯磺酸钠（DBS），阳离子表面活性剂十六烷基溴化铵（CTAB），非离子表面活性剂聚氧乙烯醚类、吐温和生物表面活性剂磷脂等。

Gittins等[55]以表面活性剂TOAB为配体，采用改进的两相法（甲苯/水）反应体系合成Pd NPs，其中TOAB起着转移和配体的双重作用。Coronado等[56]采用类似的方法合成Pd NPs粒径0.9~3.5 nm，其中表面活性剂的浓度高低和还原剂添加后的搅拌速度是影响Pd NPs粒径大小的重要因素。李冰等[57]以表面活性剂吐温20（TWEEN-20）作为钯保护剂，还原剂为$NaBH_4$，被还原的Pd^{2+}与聚氧乙烯链的氧原子形成配位，将钯原子表面包围，阻止纳米粒子的聚集，从而达到控制粒径的目的。

如图1.4所示，Reetz等[58]以表面活性剂$R_4N^+Br^-$为配体，采用电化学法合成Pd NPs，通过TEM、XRD等技术手段表征粒径为1.2~5 nm的Pd NPs，结果表明，溶剂极性、电极间距离、电流密度、电荷流量和温度等参数是影响Pd NPs的粒径的关键因素。

茹婷婷等[59]利用表面活性剂N, N-二甲基溴乙酸钠（OTAB-Na）制备Pd NPs。通过对合成条件改变，制备出具有枝化结构和凹面体的结构的Pd NPs，且2种结构的Pd NPs的粒径均一、分散性好，在经催化性能测试发现，枝化结构的Pd NPs的催化性能比凹面体结构的Pd NPs高。吕中等[60]以氯化钯为前驱体，利用4-羟乙基哌嗪乙磺（HEPES）溶液作为还原剂和配体稳定剂合成Pd NPs，运用能量分析光谱仪（EDX）、选区电子衍射

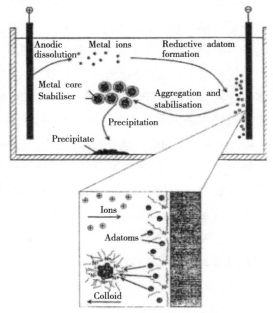

图1.4 Pd NPs合成示意图[58]

（SAED）等对所合成样品进行表征，研究表明，在添加了表面活性剂后，可以提高Pd NPs的分散性并可以减小纳米粒子的粒径。

在合成反应过程中，使用表面活性剂作为纳米颗粒的配体的主要优点，首先表面活性剂在合成过程中可以作为溶剂、还原剂和Pd NPs的配体；其次是由于它们与金属表面相互作用力较弱，在配体交换反应中可以被利用（其中表面活性剂可以被取代更强的结合配体），也可以提高钯纳米的催化作用[55]。

1.3.3　聚合物和树枝状聚合物

聚合物中含有丰富羧基和氨基的可与钯形成配合物，因此，该类聚合物可用来合成Pd NPs。Shi等[61]利用聚乙烯吡咯烷酮（PVP）为稳定剂，以UV、TEM、XPS表征后，证明合成PVP保护的Pd NPs。王艳丽等[62]采用在PVP存在下，利用还原Pd^{2+}，制备出Pd NPs，用TEM、FTIR、UV-Vis和XPS等技术手段对Pd NPs表面形貌和结构进行表征，研究表明，Pd NPs的粒径均一，平均粒径约为3 nm，所合成的Pd NPs与PVP上的羰基基团通过配位作用使Pd NPs稳定地存在且分散性好。此外，Feng等[63]通过功能化聚乙二醇（PEG）作为配体合成稳定水溶性Pd NPs，并证实通过控制配体的剂量和还原剂的种类就能对水溶性Pd NPs的大小进行控制，而且特别地证实了Pd NPs的尺寸和表面特性在影响催化特性方面扮演着重要的角色，稳定的金属Pd NPs被证实是苯甲醇氧化的活跃中心和水溶性Pd NPs催化活性也能扩展到各种醇的局部氧化。

树枝状聚合物在20世纪末期合成的一类高分子，其具有高度对称结构、高度多支化和分散性的高分子化合物，外观呈树枝状，因此类化合物具有的特殊结构可以稳定金属纳米粒子，经常用作Pd NPs的配体。树枝状聚合物中聚酰胺-胺（PAMAM）[64]和聚丙烯亚胺（PPI）[65]是合成小尺寸的Pd NPs最常用的树枝状聚合物（图1.5）。树枝状聚合物由于其清晰的官能团和结构，从而在合成过程中可以控制Pd NPs的尺寸，例如，以PAMAM聚合物而言，三级胺可作为配体与钯离子进行配位[66]。袁芳芳[67]采用化学还原法用树枝状聚合物PAMAM为配体，以氯化钯（$PdCl_2$）为前躯体，还原剂为$NaBH_4$制备PAMAM保护的Pd NPs，并用所制备的Pd NPs负载到载体上，结果表明，载体的催化剂有优异的催化性能。

G1 PAMAM树枝状聚合物　　G1 PPI 树枝状聚合物

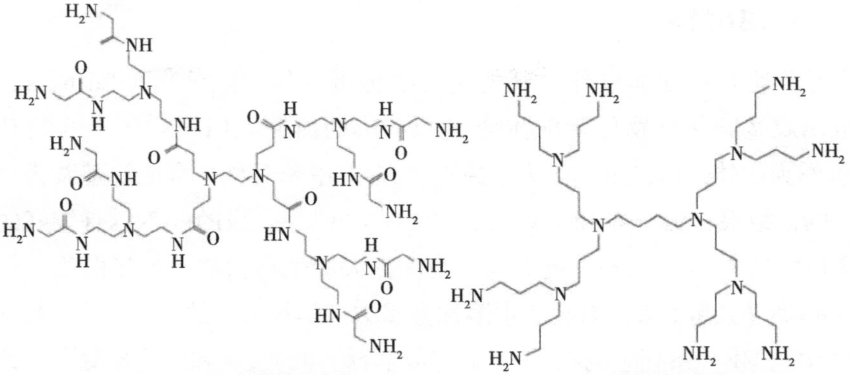

图1.5 树枝状聚合物PAMAM和PPI的结构图[64-65]

此外，介孔类聚合物也可以用于Pd NPs的合成，如图1.6所示，Xing等[68]采用功能化介孔聚合物，使用$NaBH_4$为还原剂，合成Pd NPs，并将其作为催化剂应用于Heck反应，具有良好的空气稳定性、耐久性、高活性和可重复性。

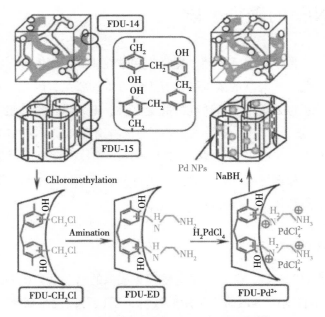

图1.6 介孔聚合物合成Pd NPs示意图[68]

1.3.4 其他配体

除了上述几种配体以外，其他化合物也可以作为配体保护Pd NPs。我们课题组用烷基胺化合物作为配体合成Pd NPs的研究。Li等[69]用十八烷基伯胺（SA）作为配体，以$NaBH_4$作为还原剂，在水-甲苯两相中合成烷基胺为稳定剂的Pd NPs，结果表明，当Pd：SA摩尔比为1：7时，合成的Pd NPs有较好的分散性，平均粒径0.8～5.6 nm。近年来，空间庞大的生物物质（如蛋白质、多肽和DNA）的使用引起了人们对控制水溶性金属纳米颗粒生长的关注[70-73]。此外，功能化Pd NPs提供一系列生物相关应用，如混合结构的组装和生物检测[74]。改变稳定肽链中的氨基酸序列对Pd NPs的粒径具有显著影响，因为多肽不仅可以作为稳定剂而且可以作为温和的还原剂起作用。Chiu等[75]以多肽既为配体又作为还原剂合成Pd NPs，结果表明，通过改变多肽中色氨酸的比例，所合成纳米粒子的粒径2.6～6.6 nm，且分散性良好。

1.4 Pd NPs的表征手段

钯纳米材料在催化、敏感特性、光吸收和磁性能等方面体现出特有的物理和化学性质，而这些性质与其的微观结构有千丝万缕的联系，因此，对钯纳米材料的性能、结构分析也是纳米学科研究的关键之一。钯纳米材料的表征和分析方法有TEM、SEM、原子力显微镜（AFM）、扫描隧道显微镜（STM）、X-射线光电子能谱（XPS）、XRD、UV-vis、FTIR、TGA等技术，可以对钯纳米的粒子大小、颗粒形貌、粒度分布和性能等方面进行分析表征，本书从以下几个方面对Pd NPs的表征手段和技术进行探讨。

1.4.1 TEM

以电子束为照明光源，通过电磁透镜聚焦成像的电子学分析表征技术，称为TEM。该技术是目前Pd NPs研究常用的技术，不仅可以用来研究Pd NPs的结晶情况，还可以直观地分析Pd NPs大小、形貌、分散性和结构等[76-77]参数。在测试时，一般是将Pd NPs通过超声作用下分散在乙醇或甲醇中，并将溶剂自然挥发，滴在专用的铜网上进行观察，常用的TEM设备有日立H-800型、H600A-2型、

JEM-2010型和200CX型等。如图1.7所示，钯纳米材料的形状通常有球形、树枝状、棒状、线状、三角形、方形、多面体等，它们的形状与Pd NPs制备方法、反应时间、配体选择等因素有密切关系[78-79]。

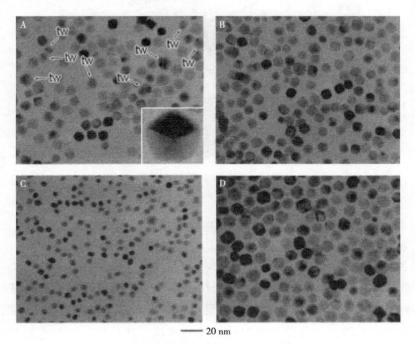

图1.7　不同反应时间Pd NPs的TEM图[78]

1.4.2　SEM

第一台商用的SEM在1965年问世，它主要是由电子光学系统、信号放大系统、显示系统等组成。SEM的成像原理不同于一般光学显微镜，是通过阴极射线管荧光屏上扫描成像，是固体样品表面分析的有效表征工具。与TEM相比SEM的优点有扫描材料图像会的立体感强，成像的范围大、分辨率也较高，制备测试样品的方法比TEM简单，还原度较高，可直接观察样品，能够真实地反映样品表面形态。因此，SEM在纳米材料研究中，经常用于观察Pd NPs的形貌、粒径大小、分散状况、成分分析等信息[80-83]。Cheng等[84]采用SEM观测溶液中钯纳米线的自组装过程，图1.8是钯纳米线的SEM图，其中2条白色的粗线是电极，与电极垂直的细线条即为所制备的钯纳米线。

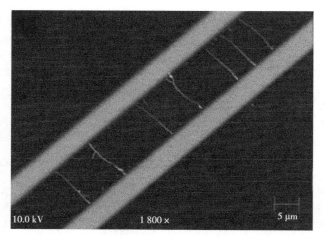

图1.8　钯纳米线的SEM图[84]

1.4.3　原子力显微镜

AFM是由1986年诺贝尔物理学奖获得者宾尼博士等发明的，此分析仪器是利用探针和测试样品表面微弱的原子间作用力的变化观察材料表面结构。AFM可以得到导体和半导体的表面结构，而且对非导体的表面结构也可以观察，弥补了其他显微镜只能观察导体和非导体结构的不足，并且其分辨率也超过了普通SEM的分辨率，可以达到原子级，而且AFM对实验工作环境及样品准备的要求要比电镜少[85-86]。

1.4.4　STM

STM的出现表明纳米技术的研究进入了一个新的阶段，其是一种基于量子隧道理论为原理的显微镜，成为Pd NPs的主要分析表征的工具之一[87-89]。STM可以实时观测原子在材料表面排列状态，以及和表明电子行为相关的物理与化学性质，STM图像具有极高的分辨率，在样品平行方向的分辨率可达0.04 nm，垂直方向的分辨率达到0.01 nm，因此，可以得到钯纳米材料表面的形貌信息和表面电子态信息。但由于STM受限于只可直接观测导体或半导体材料的表面结构，对于非导钯纳米材料必须在其表面覆盖导电层，导电层会掩盖材料表面的结构细节，从而限制了STM的使用范围。

1.4.5 UV-vis

UV-vis属于分子光谱,是材料在紫外吸收-可见光子造成分子中电子能级的跃迁而产生的吸收光谱。因为金属粒子等离子体共振激发或带间的吸收,使它们在紫外-可见区具有吸收带,并且不同的贵金属离子具有相应的特征吸收光谱[66,90],因此UV-vis是研究液态Pd NPs最常用的表征手段之一,可以初步研究粒子的粒径和形态,并且方法简便快捷,时间短,重现性好。如图1.9所示,Xiong等[91]研究发现,随着钯纳米粒径的增加,吸收峰红移近60 nm(从330 nm移至390 nm),表明Pd NPs的粒径对UV-vis特征吸收峰影响很大。因此,通过UV-vis可以得到Pd NPs粒径、结构等方面的信息。

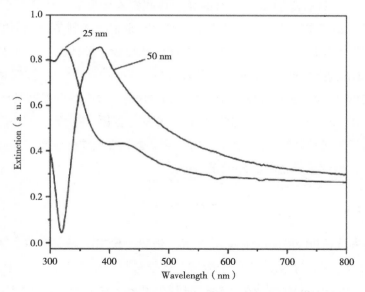

图1.9　不同粒径尺寸Pd NPs的UV-vis图[91]

1.4.6 XRD

XRD有照相法和衍射仪法2种方法。照相法应用非常广泛,一般是采用底片来记录衍射线的;衍射仪法通过采用与计算机结合,因其较高的分辨率,可对纳米颗粒进行定性分析和定量分析。XRD可获得晶体的物质组成、结构和分子间的相互作用等信息,并根据特征峰可鉴定Pd NPs的物相和含量,还可以判断粒子的尺寸大小[92]。Pd NPs的粒径大小计算依据德拜-谢勒方程(Sebye-Scherrer), d=kλ/（B cos θ），在公式中d为粒径大小，k为Scherrer常数，θ为衍射角，B为晶体粒度细化引起的宽化度。Arjona等[93]采用无任何表面活性剂和添加剂的条件下通过电化学法制备了4种花状钯纳米颗粒，经XRD分析后，钯纳米颗粒的4个特征衍射峰，2θ分别位于40.21°、46.80°、68.18°和82.28°，分别对应标准粉末衍射图谱中Pd原子（111）、（200）、（222）和（311）晶面，通过Debye-Scherrer方程计算得出，所合成的钯纳米颗粒的粒径分别为33 nm、56 nm、44 nm和47 nm。

1.4.7 FTIR

迄今为止，红外光谱仪发展已经过3代，其中FTIR是最广泛的用于表征纳米颗粒的一种，具有高通量、多通道和波数测定准确度高的优点。红外光谱对Pd NPs之间的相互作用的分析十分灵敏，可以确定钯纳米制备过程中钯原子与配体之间是否发生作用[94]。合成的Pd NPs与配体相比较，配体保护的Pd NPs的FTIR谱图出现基团的移动或者新的吸收特征峰。Yonezawa等[95]合成季铵烷基异氰化合物为配体的Pd NPs，通过FTIR表征，发现Pd NPs与—NC结合的特征峰相对于游离配体增加了约90 cm^{-1}，说明配体通过—NC基与钯原子结合。Lu等[96]首先发现Pd、Pt等铂族金属及其纳米膜在CO、CN$^-$等附着于纳米材料表面时，与吸附的金属表面相比，出现了异常的红外光谱特征。

1.4.8 XPS

XPS用来分析Pd NPs的表面化学组成、钯原子价态、表面形貌、结构状态及表面能态的分布等，X-射线光电子能谱分析仪最早是由瑞典皇家科学院院士Kai Manne Borje Siegbahn教授[34]在1954年研制出世界上第一台光电子能谱仪，其原理就是X射线激发被测样品发生光电离，从而发射光电子。XPS表征技术可用于Pd NPs表面组成的分析，可以透过配体如高分子或表面活性剂的保护层直接检测到钯原子表面，可分析Pd NPs的表面化学组成以及其表面的原子价态[97-98]。

1.4.9 TGA

TGA是一种在Pd NPs制备和检测方面常规的表征手段，一般用来研究Pd NPs的热稳定性和配体的组分，具有定量性准确，能推测出Pd NPs的质量变化和变化的速率快慢。当温度升高到一定度数时，Pd NPs会发生质量上的变化，这是由于有机配体在高温下会分解，而TGA可以提供Pd NPs中钯和有机配体的质量比，质量发生变化时所需的温度，并且TGA还可给出关于Pd NPs的热稳定方面的信息。Aymonier等[23]制备聚甲基丙烯酸甲酯（PMMA）为配体，并且合成了1.9~2.5 nm不同粒径的Pd NPs，通过TGA分析研究表明当存在少量Pd（含量为0.000 5~0.005 V/V）时，配体的分解温度延迟75℃，证明钯原子的存在明显改善了所合成Pd NPs的热稳定性。

1.5 Pd NPs的应用研究进展

1.5.1 催化剂

在工业领域中，把贵金属当作催化剂具有活性高、回收方便、可以循环使用并且具有催化效率高等优点，所有的贵金属如金、铂、钯等都可应用在催化剂领域，但由于价格昂贵使成本增加。但是，块状贵金属作为催化剂，利用率低下，而且在催化反应过程中，所产生的中间体或产物容易中毒，易导致其催化性能降低[99]。随着纳米技术的不断发展，贵金属在催化领域中的应用也越来越广，趋向于工业化。其中，贵金属钯原子与铂原子有相同的晶格结构，以及相似的原子半径和晶格能量[100-101]，最重要的是价格要比铂便宜很多。目前，全球钯开采量的一半以上都用来代替铂用作催化剂对汽车尾气的处理，因此，钯纳米材料在处理汽车尾气中起到十分关键的作用[99]。

钯作为催化剂还可以应用于有机小分子、化合物的氧化、还原及合成的反应等，其中研究比较多的是在Suzuki与Heck反应方面。2010年，Heck、根岸英一和铃木章3位科学家由于在"钯催化交叉偶联反应"研究领域所作出的杰出贡献，荣获了诺贝尔化学奖，这一技术可以让科研工作者能够精确的制造所需要的复杂化合物，目前"钯催化交叉偶联反应"技术在科研、医药生产和电子工业等领域已经得到广泛应用[102]。于克鹏[103]以树状大分子PAMAM为模板制备稳定的Pd NPs催化剂，通过催化Suzuki偶联反应检验其催化效果，研究表明此催化剂，反应条件温和，生成的产品易于分离提纯，并且产率可达96%，催化剂在第三次后，催化效率仍可达80%。

如图1.10所示，Fujii等[104]合成了聚吡咯-Pd NPs（PPy-Pd NPs），通过SEM

和TEM观察到纳米粒子为核壳形态，采用XRD表征发现，Pd^{0}存在于核壳的表面，同时，Pd NPs作为Suzuki型偶联反应形成C—C键的高效催化剂。

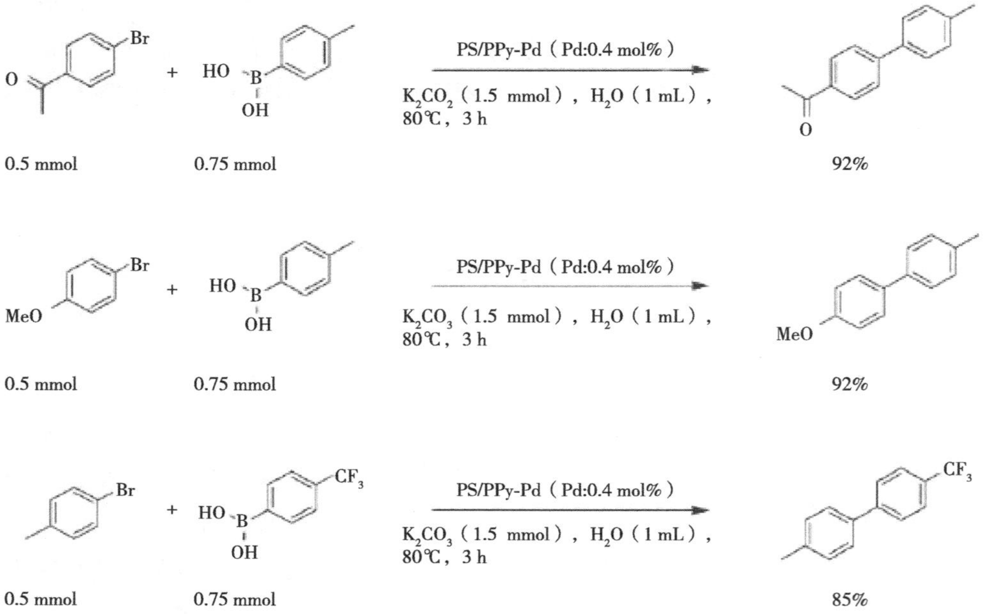

图1.10　（PPy-Pd NPs）进行Suzuki交叉偶联反应[104]

此外，Pd NPs作为催化剂在催化加氢方面也有广泛的应用[105-106]。Wilson等[107]在催化烯丙醇加氢的反应中研究了粒径在1.3～1.9 nm的Pd NPs的催化性质。经研究表明，加氢反应与催化剂钯纳米颗粒的电子和几何性质相关，而钯纳米颗粒的电子和几何性质在此粒径范围内变化很快。当Pd NPs的粒径小于1.5 nm时，加氢的动力学主要由其电子效应支配；而当Pd NPs粒径在1.5～1.9 nm时，由于几何效应决定着加氢动力学，随着Pd NPs粒径的减小，其电子性质也会从金属转变为绝缘体，这与分子的电子性质相类似[108]，正因为此种变化使得粒径小于1.5 nm的Pd NPs具有了催化性质。Yamauchi等[109]研究了粒径为（2.6±0.4）nm和（7.0±0.9）nm 2种Pd NPs的贮氢能力，研究表明，随着Pd NPs粒径的减小，其贮氢的能力也会逐渐降低，但要比钯原子的贮氢能力高1.2倍。

1.5.2　传感器

由于Pd NPs的高催化效率、高比表面积和小尺寸等优异特性，使其在传感器方面有着优良的性能，Pd NPs通过与电化学技术相结合的电化学钯纳米传感器，相对于传统检测技术手段，具有成本低、高灵敏度、选择性好、微型化和快速化等特点，在环境检测、食品检测、医学、军事等领域有着广泛的应用前景[110-113]。

Li等[114]采用甲苯和水相混合的Brust两相法，以十二烷基胺和十八烷基胺有机物为配体，合成的Pd NPs用于检测人体血液中的葡萄糖含量，研究表明，所制备的电化学传感器具有良好的稳定性、重现性和高灵敏性。

钯纳米在环境分析中也具有非常广泛的应用。由于现代工业的快速发展，环境污染问题变得越来越严重，得到了越来越多的关注，环境污染物未经处理排入江河湖泊或进入土壤，在我国大部分地区都出现并且有日益严重的趋势[115]，由于环境污染物对人体健康的毒害作用十分严重，也是全人类目前急需解决的共性问题。钯纳米材料制备的电化学传感器在环境监测中有着广泛的应用[116-118]。Sakthinathan等[119]研制了Pd NPs-β-CD复合纳米材料的电化学传感器，用在环境污染物水合肼的检测，线性范围为0.05～1 600 μm，最低检出限为28 nm，此传感器具有检出限低、选择性好和灵敏度高的优点。杨帅等[120]制备出Pd NPs/MWNTs修饰的电化学传感器，可以快速的对Cr^{6+}进行电化学测定，此方法具有选择性好、检出限低的优点。朱伟明等[121]合成了无负载的金-钯合金纳米粒子，该合成方法简便、纳米粒子的性质稳定，此纳米粒子修饰的玻碳电极催化性能好、成本低、重现性好，检出限达到8×10^{-7} mol/L，可用于快速检测食品中的H_2O_2残留量。此外，Pd NPs还可以用于生物传感器，如DNA、核酸、酶等方面的生物传感器[112,122-123]。

1.6 立题背景和研究内容

1.6.1 立题背景

在分析测定的众多技术之中，传感器因其灵敏度高、选择性好、操作简单、检测快速、成本低廉等优点，在近些年发展迅速，已成为环境分析化学、分子生物学、生物化学等学科的研究热点[124]。

随着人们对纳米技术的不断深入研究，也为纳米材料在传感器方面的应用开辟了新的思路。Pd NPs因其本身所具有表面界面效应、小尺寸效应和宏观量子隧道效应等特殊性质，继而展现出许多优异的特性，如催化效率高、比表面积大、表面活性高等，将其应用传感器的研制有着广泛的应用前景。

但是，Pd NPs的稳定性差一直是研究过程中存在的重要问题，并且大部分的Pd NPs均是非水溶性的，还有部分的Pd NPs的粒径尺寸较大，从而限制了其应用，因此制备稳定性好、粒径尺寸小、水溶性好的Pd NPs已经成为研究的发展趋势。本书在迫切需要建立制备良好性能Pd NPs的通识方法背景下开展了Pd NPs稳定性和均一性可控制备的研究。Pd NPs的应用研究是制备研究的出路，因此，选择在环境中重金属离子的检测也成为本书的研究方向之一。

近年来，重金属离子的检测技术也越来越受到了科研人员的关注。至今为止，国内外测定铜离子常用的方法主要为荧光法[125]、原子光谱法[126]、质谱法[127]、电化学方法[128]、紫外可见分光光度法[129]和化学发光法[130]等，其中电化学方法由于其分析操作较为简便且不需要造价昂贵的仪器设备，而得到了广泛应用[131]。铜离子是人类所需的微量元素之一，但铜离子通过各种途径过量地摄入人体内，会富集在人的脏器官中，由于自身无法降解，会导致人类严重的疾病的发生。此外，铜离子对环境特别是水体的污染也会造成严重的后果，在我国铜离子及其化合物被视为需优先控制的环境污染物。因此，以Pd NPs为基础，构建操作简便、分析快速、成本低的电化学传感器作为手段的铜离子检测技术具有重要的现实价值和理论意义。

1.6.2 研究内容和创新点

1.6.2.1 研究内容

本书的研究内容主要包括Pd NPs的制备和应用两大部分。首先，分别采用以NAC和DMF为配体，制备出有机小分子保护的小尺寸水溶性Pd NPs；通过UV-vis、FTIR、TEM、XPS、TGA和动态光散射（DLS）等技术对Pd NPs进行表征；其次，结合反向-高效液相色谱法（RP-HPLC）对Pd NPs进行分离纯化，从而得到Pd NPs更加丰富和具体的结构信息；最后，将水溶性Pd NPs修饰电极应用于检测环境中重金属离子，期望获得具有灵敏度高、选择性好、稳定性优良、成本低廉的Pd NPs电化学传感器。主要研究内容如下。

（1）NAC保护Pd NPs的制备与表征。在冰浴条件下，以NAC作为配体，通过$NaBH_4$还原氯钯酸合成了NAC为配体保护的水溶性Pd NPs（NAC-Pd NPs）。

通过UV-vis、XPS、FTIR、TGA、TEM和DLS对不同摩尔比合成的NAC-Pd NPs进行表征，探索制备分散性好、粒径分布均一的Pd NPs。

（2）NAC保护Pd NPs的分离。采用C18色谱柱（250 mm×4.6 mm，5 μm），流动相为甲醇和四丁基氟化铵（$Bu_4N^+F^-$）水溶液，结合梯度洗脱，利用反向离子对高效液相色谱法分离水溶性NAC-Pd NPs。研究了$Bu_4N^+F^-$和甲醇含量对NAC-Pd NPs分离的影响，利用紫外-可见光谱和基质辅助激光解析串联飞行时间质谱对所分离的Pd NPs组分进行分析。

（3）DMF保护Pd NPs的制备与表征。采用一步法合成DMF作为配体保护的水溶性Pd NPs（DMF-Pd NPs），其中DMF起着溶液、配体和还原剂三重作用。通过UV-vis、荧光光谱（FL）、FTIR、TGA、TEM和DLS对合成的DMF-Pd NPs进行表征。

（4）DMF保护Pd NPs的分离。采用C18色谱柱（250 mm×4.6 mm，5 μm），流动相为甲醇和水，结合梯度洗脱，利用RP-HPLC分离DMF-Pd NPs。研究了流动相中不同甲醇含量对DMG-Pd NPs分离的影响，利用UV-vis、FL和基质辅助激光解析串联飞行时间质谱对所分离的Pd NPs组分进行分析。

（5）Pd NPs修饰玻碳电极对Cu^{2+}的电化学检测。采用滴涂法制备DMF-Pd NPs修饰玻碳电极，并在修饰电极电化学表征的基础上，通过优化DMF-Pd NPs/GC修饰电极对Cu^{2+}的电化学响应性能，建立了一种基于DMF-Pd NPs/GC修饰电极对Cu^{2+}的测定方法。

1.6.2.2 创新点

（1）研究建立了一种有机配体稳定钯粒子的制备方法，该方法具有一定的普适性的，可以推广应用到其他有机配体或者其他贵金属纳米材料的制备过程中。

（2）首次将高效液相色谱法应用于Pd NPs组分的分离，并利用UV-vis、FL和MS测定Pd NPs组分中所包含的钯原子和配体数目，建立了一种分析Pd NPs组成的有效方法。

（3）利用所制备Pd NPs良好的溶解性和稳定性，成功研制了DMF-Pd NPs/GC修饰电极，通过对其电化学性能的研究和优化，该修饰电极具有成本低廉、灵敏度高和选择性好等优点。

参考文献

[1] 张晶. 金属离子掺杂ZnO微/纳米结构的制备及其光催化性质研究[D]. 西安: 陕西师范大学, 2011.

[2] GOESMANN H, FELDMANN C. Nanoparticulate functional materials[J]. Angewandte chemie international edition, 2010, 49(8): 1362-1395.

[3] MOORES A, GOETTMANN F. The plasmon band in noble metal nanoparticles: an introduction to theory and applications[J]. New journal of chemistry, 2006, 30(8): 1121-1132.

[4] 聂涛. 电活性PEDOT/纳米复合材料应用于B族维生素的检测[D]. 南昌: 江西农业大学, 2014.

[5] 朱世东, 周根树, 蔡锐, 等. 纳米材料国内外研究进展Ⅰ: 纳米材料的结构、特异效应与性能[J]. 热处理技术与装备, 2010, 31(3): 1-5, 26.

[6] LIU S Q, YU J H, JU H X. Renewable phenol biosensor based on a tyrosinase-colloidal gold modified carbon paste electrode[J]. Journal of electroanalytical chemistry, 2003, 540: 61-67.

[7] 董树荣, 涂江平, 王春生, 等. 纳米碳管制备的研究[J]. 材料科学与工艺, 1998(1): 31-35.

[8] GHOLINEJAD M, AHMADI J. Assemblies of copper ferrite and palladium nanoparticles on silica microparticles as a magnetically recoverable catalyst for sonogashira reaction under mild conditions[J]. Journal of chemininformatics, 2015, 46(40): 973-979.

[9] ZHAI H Z, LI J B, ZHANG S X, et al. Preparation of tetragonal zirconia containing titanium nitride powder by selective nitridation[J]. Journal of materials chemistry, 2001, 11(4): 1092-1095.

[10] CHEN S W, INGRAM R S, HOSTETLER M J, et al. Gold nanoelectrodes of varied size: transition to molecule-like charging[J]. Science, 1998, 280(5372): 2098-2101.

[11] GLAZMAN L I. Single electron tunneling[J]. Journal of low temperature physics, 2000, 118(5): 247-269.

[12] 夏和生, 王琪. 纳米技术进展[J]. 高分子材料科学与工程, 2001, 17(4): 1-6.

[13] 牛志强. 钯、铂纳米晶的调控合成与催化性能研究[D]. 北京: 清华大学, 2012.

[14] GROSS G M, GRATE J W, SYNOVEC R E. Monolayer-protected gold nanoparticles as an efficient stationary phase for open tubular gas chromatography using a square capillary: model for chip-based gas chromatography in square cornered microfabricated channels[J]. Journal of chromatography A, 2004, 1029(1): 185-192.

[15] 王焕英. 纳米材料的制备及应用研究[J]. 衡水师专学报, 2002, 4(2): 44-47.

[16] SIEGEL R W, RAMASAMY S, HAHN H, et al. Synthesis, characterization, and properties of nanophase TiO_2[J]. Journal of materials research, 1988, 3(6): 1367-1372.

[17] AYESH A I, THAKER S, QAMHIEH N. Size-controlled Pd nanocluster grown by plasma gas-condensation method[J]. Journal of nanoparticle research, 2011, 13(3): 1125-1131.

[18] 钟俊辉. 纳米粉末的制取方法[J]. 粉末冶金技术, 1995(1): 48-56.

[19] 秦聪丽. 植物还原法制备金属纳米颗粒的研究[D]. 北京: 北京服装学院, 2015.

[20] SCHÜRMANN U, TAKELE H, ZAPOROJTCHENKO V. Optical and electrical properties of polymer metal nanocomposites prepared by magnetron co-sputtering[J]. Thin solid films, 2006, 515(2): 801-804.

[21] PERRIN J, DESPAX B, KAY E. Optical properties and microstructure of gold-fluorocarbon-polymer composite films[J]. Physical review b, 1985, 32(2): 719-732.

[22] NAKAO Y. Noble metal solid sols in poly(methyl methacrylate)[J]. Journal of colloid and interface science, 1995, 171(2): 386-391.

[23] AYMONIER C, BORTZMEYER D, THOMANN R. Poly(methyl methacrylate)/palladium nanocomposites: synthesis and characterization of the morphological, thermomechanical, and thermal properties[J]. Chemistry of materials, 2003, 15(25): 4874-4878.

[24] WU X L, TAKESHITA S, TADUMI K, et al. Preparation of noble metal/polymer nanocomposites via in situ polymerization and metal complex

[25] DESHMUKH R D, COMPOSTO R J. Surface segregation and formation of silver nanoparticles created in situ in poly(methyl methacrylate) films[J]. Chemistry of materials, 2007, 19(4): 745-754.

[26] 甘颖, 徐国财, 曹震, 等. 纳米Pd/PEG复合材料棒状结构的制备与表征[J]. 化工时刊, 2012, 26(2): 1-4.

[27] 谭德新, 王艳丽, 徐国财. 纳米钯粒子的超声制备与表征[J]. 无机化学学报, 2006, 22(10): 1921-1924.

[28] 邱晓峰, 朱俊杰. 超声化学制备单分散金属纳米钯[J]. 无机化学学报, 2003, 19(7): 766-770.

[29] TU W X, LIU H F. Continuous synthesis of colloidal metal nanoclusters by microwave irradiation[J]. Chemistry of materials, 2000, 12(2): 564-567.

[30] ZHU Y J, CHEN F. Microwave-assisted preparation of inorganic nanostructures in liquid phase[J]. Chemical reviews, 2014, 114(12): 6462-6555.

[31] 刘华蓉, 张志成, 钱逸泰, 等. γ-射线辐照法制备纳米Cu-Pd合金粉末的影响因素[J]. 无机化学学报, 1999, 15(3): 388-392.

[32] GROSS M E, APPELBAUM A, GALLAGHER P K. Laser direct-write metallization in thin palladium acetate films[J]. Journal of applied physics, 1987, 61(4): 1628-1632.

[33] SHAUKAT M S, ZULFIQAR S, SARWAR M I. Incorporation of palladium nanoparticles into aromatic polyamide/clay nanocomposites through facile dry route[J]. Polymer science series B, 2015, 57(4): 380-386.

[34] HAN G Y, GUO B, ZHANG L W, et al. Conductive gold films assembled on electrospun poly(methyl methacrylate) fibrous mats[J]. Advanced materials, 2006, 18(13): 1709-1712.

[35] LIU Z L, FENG Y H, WU X F, et al. Preparation and enhanced electrocatalytic activity of graphene supported palladium nanoparticles with multi-edges and corners[J]. RSC advances, 2016, 6(101): 98708-98716.

[36] HACHE F, RICARD D, FLYTZANIS C. Optical nonlinearities of small metal particles: surface-mediated resonance and quantum size effects[J]. Journal of the optical society of America B, 1986, 3(12): 1647-1655.

[37] 杨建新, 王寅, 毛玉荣, 等. Pd/C纳米催化剂的制备及催化Heck反应研究[J]. 化工新型材料, 2014, 42(4): 132-135.

[38] VIET L N, DUC C N, HIROHITO H, et al. Chemical synthesis and characterization of palladium nanoparticles[J]. Advances in natural sciences: nanoscience and nanotechnology, 2010, 1(3): 2043-2047.

[39] CHAN Y NG CHEONG, CRAIG G S W, SCHROCK R R, et al. Synthesis of palladium and platinum nanoclusters within microphase-separated diblock copolymers[J]. Chemistry of materials, 1992, 4(4): 885-894.

[40] KRÁLIK M, HRONEC M, LORA S, et al. Microporous poly-N,N-dimethylacrylamide-p-styrylsulfonate-methylene bis(acrylamide): a promising support for metal catalysis[J]. Journal of molecular catalysis A: chemical, 1995, 97(3): 145-155.

[41] 雷彬, 张旭, 谭文松, 等. Klebsiella Pneumoniae ECU-15生物纳米钯催化剂的制备、表征及其合成机理初探[J]. 高校化学工程学报, 2015, 29(4): 873-880.

[42] MOHAMMADINEJAD R, KARIMI S, IRAVANI S, et al. Plant-derived nanostructures: types and applications[J]. Green chemistry, 2016, 18(1): 20-52.

[43] NADAGOUDA M N, VARMA R S. Green synthesis of silver and palladium nanoparticles at room temperature using coffee and tea extract[J]. Green chemistry, 2008, 10(8): 859-862.

[44] AKHTAR M S, PANWAR J, YUN Y S. Biogenic synthesis of metallic nanoparticles by plant extracts[J]. ACS sustainable chemistry & engineering, 2013, 1(6): 591-602.

[45] DAUTHAL P, MUKHOPADHYAY M. Biosynthesis of palladium nanoparticles using delonix regia leaf extract and its catalytic activity for nitro-aromatics hydrogenation[J]. Industrial & engineering chemistry research, 2013, 52(51): 18131-18139.

[46] 贺媛媛, 傅吉全. 植物还原法制备钯纳米单质的定量研究[J]. 石油炼制与化工, 2018(4): 101-104.

[47] 秦聪丽, 傅吉全. 桑叶提取物还原制备钯纳米颗粒[J]. 贵金属, 2016, 37(3): 29-32.

[48] 秦聪丽, 傅吉全. 植物法制备钯纳米颗粒及其应用[J]. 工业催化, 2016, 24(3):

58-61.

[49] COOKSON J. The preparation of palladium nanoparticles[J]. Platinum metals review, 2012, 56(2): 83-98.

[50] BRUST M, FINK J, BETHELL D, et al. Synthesis and reactions of functionalised gold nanoparticles[J]. Journal of the chemical society, chemical communications, 1995(16): 1655-1656.

[51] CHEN S W, HUANG K, STEARNS J A. Alkanethiolate-protected palladium nanoparticles[J]. Chemistry of materials, 2000, 12(2): 540-547.

[52] SADEGHMOGHADDAM E, LAM C, CHOI D, et al. Synthesis and catalytic properties of alkanethiolate-capped Pd nanoparticles generated from sodium S-dodecylthiosulfate[J]. Journal of materials chemistry, 2011, 21(2): 307-312.

[53] ZAMBORINI F P, GROSS S M, MURRAY R W. Synthesis, characterization, reactivity, and electrochemistry of palladium monolayer protected clusters[J]. Langmuir, 2001, 17(2): 481-488.

[54] SAKAI N, TATSUMA T. One-step synthesis of glutathione-protected metal (Au, Ag, Cu, Pd, and Pt) cluster powders[J]. Journal materials chemistry A, 2013, 1(19): 5915-5922.

[55] GITTINS D I., CARUSO F. Spontaneous phase transfer of nanoparticulate metals from organic to aqueous media[J]. Angewandte chemie international edition, 2001, 40(16): 3001-3004.

[56] CORONADO E, RIBERA A, GARCÍA-MARTÍNEZ J, et al. Synthesis, characterization and magnetism of monodispersed water soluble palladium nanoparticles[J]. Journal of materials chemistry, 2008, 18(46): 5682-5688.

[57] 李冰. 多孔炭材料负载纳米钯催化剂的制备及催化Suzuki反应的研究[D]. 哈尔滨: 黑龙江大学, 2016.

[58] REETZ M T, WINTER M, BREINBAUER R, et al. Size-selective electrochemical preparation of surfactant-stabilized Pd-, Ni-and Pt/Pd colloids[J]. Chemistry-a european journal, 2001, 7(5): 1084-1094.

[59] 茹婷婷, 初学峰, 石莹岩, 等. 钯纳米粒子的形貌可控合成与催化性能[J]. 无机化学学报, 2017, 33(10): 1835-1842.

[60] 吕中, 张伟, 赵慧平, 等. HEPES溶液中钯纳米颗粒的制备及对Suzuki反应的

催化[J]. 华中师范大学学报(自然科学版), 2013, 47(2): 221-224.

[61] SHI W, LIU C X, SONG Y L, et al. An ascorbic acid amperometric sensor using over-oxidized polypyrrole and palladium nanoparticles composites[J]. Biosensors and bioelectronics, 2012, 38(1): 100-106.

[62] 王艳丽, 谭德新, 徐国财, 等. 超细纳米Pd粒子与聚乙烯吡咯烷酮的相互作用[J]. 高等学校化学学报, 2010, 31(5): 881-884.

[63] FENG B, HOU Z S, YANG H M, et al. Functionalized poly(ethylene glycol)-stabilized water-soluble palladium nanoparticles: property/activity relationship for the aerobic alcohol oxidation in water[J]. Langmuir, 2010, 26(4): 2505-2513.

[64] BALOGH L, TOMALIA D A. Poly(amidoamine) dendrimer-templated nanocomposites. 1. synthesis of zerovalent copper nanoclusters[J]. Journal of the American chemical society, 1998, 120(29): 7355-7356.

[65] YEUNG L K, CROOKS R M. Heck heterocoupling within a dendritic nanoreactor[J]. Nano letters, 2001, 1(1): 14-17.

[66] BRONSTEIN L M, SHIFRINA Z B. Dendrimers as encapsulating, stabilizing, or directing agents for inorganic nanoparticles[J]. Chemical reviews, 2011, 111(9): 5301-5344.

[67] 袁芳芳. 钯纳米颗粒的可控合成及其催化性能研究[D]. 武汉: 中南民族大学, 2011.

[68] XING R, LIU Y M, WU H H, et al. Preparation of active and robust palladium nanoparticle catalysts stabilized by diamine-functionalized mesoporous polymers[J]. Chemical communications, 2008(47): 6297-6299.

[69] LI Z P, GAO J, XING X T, et al. Synthesis and characterization of n-alkylamine-stabilized palladium nanoparticles for electrochemical oxidation of methane[J]. Journal physical chemistry C, 2009, 114(2): 723-733.

[70] PACARDO D B, SETHI M, JONES S E, et al. Biomimetic synthesis of Pd nanocatalysts for the stille coupling reaction[J]. ACS Nano, 2009, 3(5): 1288-1296.

[71] KRAMER R M, LI C, CARTER D C, et al. Engineered protein cages for nanomaterial synthesis[J]. Journal of the american chemical society, 2004, 126(41): 13282-13286.

[72] LI Y J, WHYBURN G P, HUANG Y. Specific peptide regulated synthesis of ultrasmall platinum nanocrystals[J]. Journal of the American chemical society, 2009, 131(44): 15998-15999.

[73] COPPAGE R, SLOCIK J M, SETHI M, et al. Elucidation of peptide effects that control the activity of nanoparticles[J]. Angewandte chemie international edition, 2010, 49(22): 3767-3770.

[74] SLOCIK J M, STONE M O, NAIK R R. Synthesis of gold nanoparticles using multifunctional peptides[J]. Small, 2005, 1(11): 1048-1052.

[75] CHIU C L, LI Y J, YU H. Size-controlled synthesis of Pd nanocrystals using a specific multifunctional peptide[J]. Nanoscale, 2010, 6(2): 927-930.

[76] GIERSIG M, MULVANEY P. Preparation of ordered colloid monolayers by electrophoretic deposition[J]. Langmuir, 1993, 9(12): 3408-3413.

[77] GOPIDAS K R, WHITESELL J K, FOX M A. Synthesis, characterization, and catalytic applications of a palladium-nanoparticle-cored dendrimer[J]. Nano letters, 2003, 3(12): 1757-1760.

[78] XIONG Y, XIA Y. Shape-controlled synthesis of metal nanostructures: the case of palladium[J]. Advanced materials, 2007, 19(20): 3385-3391.

[79] XIONG Y J, CHEN J Y, WILEY B, et al. Understanding the role of oxidative etching in the polyol synthesis of Pd nanoparticles with uniform shape and size[J]. Journal of the American chemical society, 2005, 127(20): 7332-7333.

[80] LIU R J, CROZIER P A, SMITH C M, et al. In situ electron microscopy studies of the sintering of palladium nanoparticles on alumina during catalyst regeneration processes[J]. Microsc microanal, 2004, 10(1): 77-85.

[81] JEON J H, LIM J H, KIM K M. Hybrid nanocomposites of palladium nanoparticles having POSS and MWNTs via ionic interactions[J]. Macromolecular research, 2009, 17(12): 987-994.

[82] VANCOVÁ M, SLOUF M, LANGHANS J, et al. Application of colloidal palladium nanoparticles for labeling in electron microscopy[J]. Microsc microanal, 2011, 17(5): 810-816.

[83] 韦贻春, 余会成, 李浩, 等. 钯纳米粒子修饰的三甲氧苄啶分子印迹膜传感器的制备及其识别性能研究[J]. 分析化学, 2017, 45(9): 1367-1374.

[84] CHENG C D, GONELA R K, GU Q, et al. Self-assembly of metallic nanowires from aqueous solution[J]. Nano letters, 2005, 5(1): 175-178.

[85] ESFANDIAR A, GHASEMI S, IRAJIZAD A, et al. The decoration of TiO_2/ reduced graphene oxide by Pd and Pt nanoparticles for hydrogen gas sensing[J]. International journal of hydrogen energy, 2012, 37(20): 15423-15432.

[86] LIN H L, YANG J M, LIU J Y, et al. Properties of Pd nanoparticles-embedded polyaniline multilayer film and its electrocatalytic activity for hydrazine oxidation[J]. Electrochimica acta, 2013, 90: 382-392.

[87] LU W, WANG B, WANG K D, et al. Synthesis and characterization of crystalline and amorphous palladium nanoparticles[J]. Langmuir, 2003, 19(14): 5887-5891.

[88] CORTHEY G, RUBERT A A, PICONE A L, et al. New insights into the chemistry of thiolate-protected palladium nanoparticles[J]. The journal of physical chemistry C, 2012, 116(17): 9830-9837.

[89] CORTHEY G, OLMOS-ASAR J A, CASILLAS G, et al. Influence of capping on the atomistic arrangement in palladium nanoparticles at room temperature[J]. The journal of physical chemistry C, 2014, 118(42): 24641-24647.

[90] AHMAD A, WEI Y, SYED F, et al. Size dependent catalytic activities of green synthesized gold nanoparticles and electro-catalytic oxidation of catechol on gold nanoparticles modified electrode[J]. RSC advances, 2015, 5(120): 99364-99377.

[91] XIONG Y J, CHEN J Y, WILEY B, et al. Size-dependence of surface plasmon resonance and oxidation for Pd nanocubes synthesized via a seed etching process[J]. Nano letters, 2005, 5(7): 1237-1242.

[92] CHEN M, FALKNER J, GUO W HA, et al. Synthesis and self-organization of soluble monodisperse palladium nanoclusters[J]. Journal of colloid and interface science, 2005, 287(1): 146-151.

[93] ARJONA N, GUERRA-BALCAZAR M, CUEVAS-MUNIZ F M, et al. Electrochemical synthesis of flower-like Pd nanoparticles with high tolerance toward formic acid electrooxidation[J]. RSC advances, 2013, 3(36): 15727-15733.

[94] DI GREGORIO F, BISSON L, ARMAROLI T, et al. Characterization of well

faceted palladium nanoparticles supported on alumina by transmission electron microscopy and FT-IR spectroscopy of CO adsorption[J]. Applied catalysis A: general, 2009, 352(1): 50-60.

[95] YONEZAWA T, IMAMURA K, KIMIZUKA N. Direct preparation and size control of palladium nanoparticle hydrosols by water-soluble isocyanide ligands[J]. Langmuir, 2001, 17(16): 4701-4703.

[96] LU G Q, SUN S G, CHEN S P, et al. Novel properties of dispersed Pt and Pd thin layers supported on GC for CO adsorption studied using in situ MS-FTIR reflection spectroscopy[J]. Journal of electroanalytical chemistry, 1997, 421(1): 19-23.

[97] VOOGT E H, MENS A J M, GIJZEMAN O L J, et al. XPS analysis of palladium oxide layers and particles[J]. Surface science, 1996, 350(1): 21-31.

[98] KIDAMBI S, BRUENING M L. Multilayered polyelectrolyte films containing palladium nanoparticles: synthesis, characterization, and application in selective hydrogenation[J]. Chemistry of materials, 2005, 17(2): 301-307.

[99] 左盼盼, 廉园园, 于浩, 等. 纳米钯修饰电极在碱性条件下对过氧化氢的测定[J]. 分析测试学报, 2010, 29(5): 484-487, 492.

[100] ZHANG J T, HUANG M H, MA H Y, et al. High catalytic activity of nanostructured Pd thin films electrochemically deposited on polycrystalline Pt and Au substrates towards electro-oxidation of methanol[J]. Electrochemistry communications, 2007, 9(6): 1298-1304.

[101] ATTARD G A, PRICE R, AL-AKL A. Palladium adsorption on Pt(111): a combined electrochemical and ultra-high vacuum study[J]. Electrochimica acta, 1994, 39(11): 1525-1530.

[102] ASTRUC D. The 2010 Chemistry Nobel Prize to R. F. Heck, E. Negishi, and A. Suzuki for palladium-catalyzed cross-coupling reactions[J]. Analytical and bioanalytical chemistry, 2011, 399(5): 1811-1814.

[103] 于克鹏. 聚酰胺-胺为模板的钯纳米簇催化的Suzuki偶联反应[J]. 现代化工, 2012, 32(3): 48-51.

[104] FUJII S, MATSUZAWA S, NAKAMURA Y, et al. Synthesis and characterization of polypyrrole-palladium nanocomposite-coated latex particles

and their use as a catalyst for suzuki coupling reaction in aqueous media[J]. Langmuir, 2010, 26(9): 6230-6239.

[105] KOBAYASHI H, YAMAUCHI M, KITAGAWA H, et al. Hydrogen Absorption in the core/shell interface of Pd/Pt nanoparticles[J]. Journal of the American chemical society, 2008, 130(6): 1818-1819.

[106] CRESPO E A, RUDA M, RAMOS D D S, et al. Hydrogen absorption in Pd nanoparticles of different shapes[J]. International journal of hydrogen energy, 2012, 37(19): 14831-14837.

[107] WILSON O M, KNECHT M R, GARCIA-MARTINEZ J C, et al. Effect of Pd nanoparticle size on the catalytic hydrogenation of allyl alcohol[J]. Journal of the American chemical society, 2006, 128(14): 4510-4511.

[108] GUO R, MURRAY R W. Substituent effects on redox potentials and optical gap energies of molecule-like $Au_{38}(SPhX)_{24}$ nanoparticles[J]. Journal of the American chemical society, 2005, 127(34): 12140-12143.

[109] YAMAUCHI M, IKEDA R, KITAGAWA H, et al. Nanosize effects on hydrogen storage in palladium[J]. The journal of physical chemistry C, 2008, 112(9): 3294-3299.

[110] MIJOWSKA E, ONYSZKO M, URBAS K, et al. Palladium nanoparticles deposited on graphene and its electrochemical performance for glucose sensing[J]. Applied surface science, 2015, 355: 587-592.

[111] LIM S H, WEI J, LIN J Y, et al. A glucose biosensor based on electrodeposition of palladium nanoparticles and glucose oxidase onto nafion-solubilized carbon nanotube electrode[J]. Biosens bioelectron, 2005, 20(11): 2341-2346.

[112] WANG Y H, HUANG K J, WU X. Recent advances in transition-metal dichalcogenides based electrochemical biosensors: a review[J]. Biosensors and bioelectronics, 2017, 97(supplement C): 305-316.

[113] CHENG W, COMPTON R G. Electrochemical detection of nanoparticles by 'nano-impact' methods[J]. TrAC Trends in analytical chemistry, 2014, 58(supplement C): 79-89.

[114] LI Z P, WANG X, WEN G M, et al. Application of hydrophobic palladium nanoparticles for the development of electrochemical glucose biosensor[J].

Biosensors and bioelectronics, 2011, 26(11): 4619-4623.

[115] 钟燕, 周洁丹, 刘英菊, 等. 新型巯基化合物自组装膜修饰电极对铜离子的检测[J]. 分析测试学报, 2009, 28(9): 1031-1034, 1039.

[116] ZHAO J Y, ZHU M N, ZHENG M, et al. Electrocatalytic oxidation and detection of hydrazine at carbon nanotube-supported palladium nanoparticles in strong acidic solution conditions[J]. Electrochimica acta, 2011, 56(13): 4930-4936.

[117] YI Y Z, JIANG S JN, SAHAYAM A C. Palladium nanoparticles as the modifier for the determination of Zn, As, Cd, Sb, Hg and Pb in biological samples by ultrasonic slurry sampling electrothermal vaporization inductively coupled plasma mass spectrometry[J]. Journal of analytical atomic spectrometry, 2012, 27(3): 426-431.

[118] KUMAR D P V, VEERAMANI V, CHEN S M, et al. Palladium nanoparticle incorporated porous activated carbon: electrochemical detection of toxic metal ions[J]. ACS Applied materials & interfaces, 2016, 8(2): 1319-1326.

[119] SAKTHINATHAN S, KUBENDHIRAN S, CHEN S M, et al. Functionalization of reduced graphene oxide with β-cyclodextrin modified palladium nanoparticles for the detection of hydrazine in environmental water samples[J]. Electroanalysis, 2017, 29(2): 587-594.

[120] 杨帅, 汤婉鑫, 张超, 等. PdNPs/MWNTs修饰玻碳电极的制备及其对六价铬的电化学测定[J]. 上海师范大学学报(自然科学版), 2014, 43(6): 594-599.

[121] 朱伟明, 梁新义, 庞广昌, 等. 基于金钯纳米合金修饰的过氧化氢传感器的研制[J]. 食品科学, 2012, 33(10): 311-314.

[122] WANG J, XU D, KAWDE A N, et al. Metal nanoparticle-based electrochemical stripping potentiometric detection of DNA hybridization[J]. Analytical chemistry, 2001, 73: 5576.

[123] PARK S J, TATON T A, MIRKIN C A. Array-based electrical detection of DNA with nanoparticle probes[J]. Science, 2002, 295: 1503-1506.

[124] 张彦. 巯基化合物修饰金纳米粒子的制备及其分析应用[D]. 太原: 山西大学, 2009.

[125] ZHU A W, QU Q, SHAO X L, et al. Carbon-dot-based dual-emission

nanohybrid produces a ratiometric fluorescent sensor for in vivo imaging of cellular copper ions[J]. Angewandte chemie international edition, 2012, 51(29): 7185-7189.

[126] SHAO X L, GU H, WANG Z, et al. Highly selective electrochemical strategy for monitoring of cerebral Cu^{2+} based on a carbon dot-TPEA hybridized surface[J]. Analytical chemistry, 2013, 85(1): 418-425.

[127] CHAI X, ZHOU X, ZHU A, et al. A two-channel ratiometric electrochemical biosensor for in vivo monitoring of copper ions in a rat brain using gold truncated octahedral microcages[J]. Angewandte chemie, 2013, 52(31): 8129-8133.

[128] 陈红任, 孙和鑫, 朱春城. 氧化石墨烯修饰碳糊电极循环伏安法测定铜离子[J]. 人工晶体学报, 2016(11): 2634-2638.

[129] 李志文, 李海波, 万浩, 等. 水环境重金属铜离子光学检测仪器的设计[J]. 传感技术学报, 2016(1): 146-149.

[130] 屈颖娟, 翟云会, 王亚妮. 水中铜离子的高选择性检测方法[J]. 光谱实验室, 2013(5): 2283-2286.

[131] 张磊, 王旭, 李忠平, 等. 钯纳米粒子修饰的玻碳电极对Cu^{2+}的电化学检测[J]. 山西大学学报(自然科学版), 2018, 41(4): 781-786.

第2章

NAC保护的Pd NPs的制备和表征

2.1 引言

近年来，粒径在1～100 nm贵金属纳米粒子由于其特有的物理和化学性质，引起人们极大的兴趣和关注[1-3]。贵金属纳米粒子已经在许多领域被广泛地研究和应用，例如催化[4-5]、传感器[6-7]、纳米电子学[8]、光谱学[9]和生物科学[10]等方面。贵金属纳米粒子的性质与纳米粒子的形状、尺寸大小和结构组成有着紧密的关系，例如，人们可以通过控制Pd NPs的形貌和尺寸来优化它的催化性能[11]。因此，纳米材料的制备技术也得到深入的研究。

Cookson等[12]对Pd NPs的制备方法进行了综述，对Pd NPs合成方法目前的研究主要是关注对Pd NPs分散性、粒径和形貌的控制。目前，Pd NPs的合成方法有物理法、化学法和生物法。物理法制备的Pd NPs的粒径较大、反应条件要求高、重现性差；生物法制备Pd NPs的官能团作用机制还不明确，制备技术还处于实验室阶段；与物理法和生物法相比，通过化学还原法制备的Pd NPs形貌可控、不需要复杂的仪器设备，并且可以实现纳米粒子大产量的合成[13]。在使用化学还原法制备Pd NPs时，常用的还原剂有$NaBH_4$[14]、抗坏血酸[15]、柠檬酸钠[16]、乙二醇[17]等，不同还原剂的使用，会影响到Pd NPs的粒径、形貌[18]。不同有机配体保护合成的纳米粒子的性质也会不同，其中硫醇化合物作为金属纳米粒子的配体，最早报道是Giersig等[19]用烷基硫醇作为金纳米粒子的配体被使用的，由于其所合成的纳米粒子粒径尺寸小、分散性好、稳定性好且可以在有机试剂中重复的溶解和分离，并不会发生不可逆的分解和聚合，从而也被应用于合成过程中Pd NPs的配体。Sharma等[20]合成了GSH为配体保护的Pd NPs，使用UV-vis、TEM和XPS等技术手段对Pd NPs进行表征，结果表明，所合成的Pd NPs粒径在1～4 nm，分散性好，且Pd NPs表现出超顺磁性。Gavia等[21]利用ω-羧基-S-烷硫基硫酸钠作为配体，实现了稳定的水溶性Pd NPs的合成。通过TEM、TGA、核磁共振氢谱（^1H-NMR）、UV-vis和FTIR光谱表征合成的Pd NPs，研究了改变配体链长度和还原剂$NaBH_4$的浓度，对合成Pd NPs粒径和性质的影响；同时还将合成的水溶性Pd NPs用于催化烯丙醇的加氢反应中，研究表明，Pd NPs的催化活性和选择性受到Pd NPs的表面覆盖的配体和配体链长度的影响。Sakai等[22]和Cargnello等[23]也分别研究了合成不同烷基硫醇基保护的Pd NPs，结果表明，反应时间、巯基配体与钯离子的摩尔比、还原剂的添加量影响着所合成Pd NPs的平均粒径、分散性和性质。

本章以NAC作为保护Pd NPs的配体，在冰浴条和不同NAC/Pd摩尔比条件下，通过$NaBH_4$还原H_2PdCl_2合成了水溶性Pd NPs（NAC-Pd NPs），通过UV-vis、XPS、FTIR、TGA、TEM和DLS对不同NAC/Pd摩尔比合成的NAC-Pd NPs进行表征。

2.2 实验部分

2.2.1 主要试剂和仪器

氯化钯（$PdCl_2$，>99.9%）购于Aldrich公司（Milwaukee，WI，USA）；盐酸（HCl，优级纯，北京化工厂）；NAC（>99%）、$NaBH_4$（>98%）；析膜购于北京索莱宝科技有限公司；色谱级甲醇购于Bangkok公司；实验用水为超纯水（Mill-Q Advantage A10超纯水机，默克密理博）；其他所有试剂均为分析纯及以上，无须再纯化。

ME 204万分之一天平（METLER TLIEDO，上海）；RE 52A真空旋转蒸发仪（上海亚荣生化仪器厂）；Free Zone 6真空冷冻干燥机（Labconco，美国）；Varian Cary 300紫外-可见分光光度计（Varian，Palo Alto，CA）；Perkin-Elmer Paragon 100 FTIR红外光谱仪（Birmingham，NJ）；JEOL JEM-1011透射电子显微镜（日本电子光学公司）；Leybold Heraeus SKL-12X射线光电子能谱仪（X-ray photoelectron spectroscopy，XPS）（中国）；Perkin-Elmer TGA 6热重分析仪（Waltham，MA）；Nano-ZS90 Zetasizer动态光散射分析仪（英国马尔文公司）。

2.2.2 NAC-Pd NPs的制备

Pd NPs合成具体步骤如下：称取0.442 5 g的$PdCl_2$溶于2 mol/L的HCl溶液中，并定容至5 mL的容量瓶中，每次取0.5 mL其中含有0.25 mmol的Pd^{2+}；称取2.04 g NAC溶于水中，并定容于20 mL的容量瓶中，按照NAC/Pd摩尔比为1∶4、1∶2、1∶1、2∶1、4∶1的要求取相应体积的NAC溶液。在冰浴条件下，分别用移液枪取0.1 mL、0.2 mL、0.3 mL、0.4 mL、0.8 mL、1.6 mL的NAC溶液加入含有47.5 mL的甲醇的100 mL的圆底烧瓶中，逐滴加入0.5 mL的H_2PdCl_2溶液，强烈搅拌5 min后，再逐滴加入0.2 mol/L 12.5 mL的$NaBH_4$溶液，剧烈搅拌1 h后，旋转蒸发多余的甲醇溶液,将生成的沉淀物经截留分子量为3 500 Da的透析袋透析3 d后,经真空冷冻干燥后,得到粉末状不同NAC/Pd摩尔比的NAC-Pd NPs,将制备好的Pd NPs存于4℃冰箱中备用。

2.2.3　UV-vis的测量

移取一定浓度的NAC-Pd NPs水溶液至比色皿中,空白为水溶液,用Varian Cary 300紫外-可见分光光度计扫描200~500 nm内的吸收光谱。

2.2.4　FTIR的表征

取经真空冷冻干燥处理的NAC-Pd NPs粉末样品并加入干燥的KBr粉末,在红外干燥灯下研磨均匀,然后通过压力控制机制成压片,测试波长范围为4 000~500 cm^{-1},通过Perkin-Elmer Paragon 100 FTIR红外光谱仪进行红外表征。

2.2.5　TEM的表征

通过TEM对合成的NAC-Pd NPs的分散情况和表面状态进行观察,将浓缩后的NAC-Pd NPs甲醇溶液,超声分散20 min,用移液枪将其滴至涂覆了碳膜的铜网上(300目),待溶剂自然挥发后,得到样品。TEM测定的操作电压为200 kV,通过对TEM中的纳米粒子进行测量统计,得到Pd NPs的大小尺寸及粒径分布。

2.2.6　XPS的测定

在室温下,通过SKL-12X射线光电子能谱仪对NAC-Pd NPs进行XPS测定,其中激发源为单色MgKα射线(1 253.6 eV),工作电压为10 kV,保持测量室真空度为6.7×10^{-5} Pa,将NAC-Pd NPs粉末样品用透明双面胶带粘于样品台上,测定其能谱图。数据由Keatos公司Vision转换为VAMAS格式后,使用CasaXPS 2.3.16软件进行拟合处理。

2.2.7　TGA的测定

TGA在Perkin-Elmer TGA 6热重分析仪上测定,在氮气氛围下,称取约

3.0 mg的NAC-Pd NPs，置于热重分析仪的坩埚内，以10℃/min升温速度，在20～600℃范围下，记录其失重比例和温度变化。

2.3 结果与讨论

2.3.1 NAC-Pd NPs的合成

以NAC为配体，在冰浴条件下，通过NaBH$_4$还原H$_2$PdCl$_2$合成NAC-Pd NPs，合成示意图如图2.1所示。当NAC和H$_2$PdCl$_2$溶液中加入还原剂NaBH$_4$时，由于配体NAC中的—S基极易与钯原子结合生成稳定的Pd—S键，所以NaBH$_4$还原H$_2$PdCl$_2$成为钯原子的同时，NAC也迅速与钯原子反应，将其包裹起来，阻止已生成Pd NPs的再聚集，起到了稳定和保护Pd NPs的作用。通过不同NAC/Pd摩尔比条件，分别合成1∶4、1∶2、1∶1、2∶1、4∶1的Pd NPs。当[NAC]/[Pd]=1∶4时，NAC-Pd NPs为黑色粉末，但不溶于水和甲醇。当[NAC]/[Pd]=1∶2和1∶1时，NAC-Pd NPs为黑色粉末，易溶于水和甲醇，但是NAC/Pd摩尔比为1∶2合成的Pd NPs，在水溶液中不稳定，在第4天又重新形成沉淀。当[NAC]/[Pd]=2∶1和4∶1时，合成的Pd NPs为深棕色，易溶于水和甲醇。

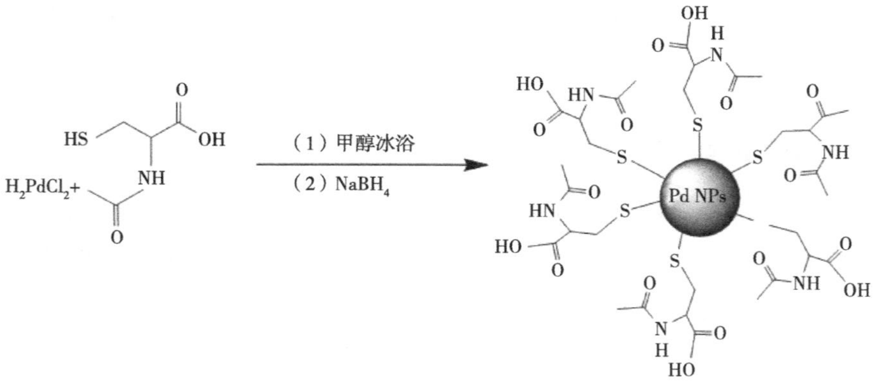

图2.1 NAC-Pd NPs的合成示意图

2.3.2 NAC-Pd NPs的UV-vis

UV-vis是用于研究贵金属纳米粒子性质的主要方法之一，可以用来反映贵金属原子被还原为贵金属纳米粒子的过程[24-25]。图2.2为NAC/Pd摩尔比分别为1∶2、1∶1、2∶1、4∶1条件下合成的NAC-Pd NPs溶液和H_2PdCl_4溶液的紫外光谱。在图中可以看出，H_2PdCl_4溶液在300 nm和400 nm有2个宽的波带，在206 nm和237 nm附近有2个特征吸收峰，是由于配体向金属电荷的转移[26-27]。NAC/Pd摩尔比为2∶1和4∶1的Pd NPs样品，在206 nm和237 nm的尖峰消失，而新产生了265 nm、325 nm和370 nm的吸收带，新吸收带的产生，是由于巯基和Pd^{II}-NAC形成的复合物[28-31]，证明并没有生成Pd^0的纳米粒子。NAC/Pd摩尔比为1∶2和1∶1的Pd NPs样品的UV-vis随着波长的增加，吸收度值在逐渐减小，而且在NAC-Pd NPs的吸收光谱没有表面等离子体共振带（SPR）的出现[32-33]，证明我们合成了NAC保护的Pd NPs。

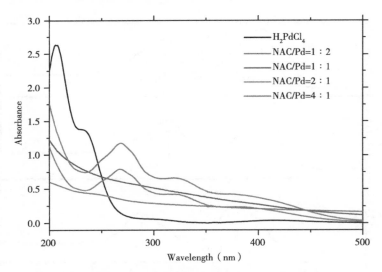

图2.2　不同摩尔比合成的NAC-Pd NPs和H_2PdCl_4的UV-vis图

2.3.3　NAC-Pd NPs的X射线光电子能谱

XPS是一种有效的手段用来分析固体样品表面的组成和化学状态的分析，被广泛用于元素分析、化合物结构的鉴定等方面，在各类材料的研究和应用中起着十分重要的作用。从XPS谱图的变化来研究贵金属纳米粒子，通过$2p_{1/2}$和$2p_{3/2}$、$3d_{3/2}$和$3d_{5/2}$等分裂线的谱线形状、能级间距和能量位置对其化合态进行鉴定，因此，XPS可以用来确定我们所制备Pd NPs的元素组成和价态，特别是活性贵金属纳米粒子的稳定性[34]。图2.3为不同摩尔比合成的NAC-Pd NPs和Pd原子的XPS图谱，其中A为摩尔比[NAC]/[Pd]=1∶1合成的Pd NPs，B为Pd原子，C为摩尔比[NAC]/[Pd]=2∶1合成的Pd NPs。从图中可以看出，摩尔比[NAC]/[Pd]=1∶1合成的Pd NPs，Pd $3d_{3/2}$和Pd $3d_{5/2}$的电子结合能分别为335.7 eV和340.9 eV，峰间距为5.20 eV，其电子结合能与Pd^0的电子结合能（Pd $3d_{3/2}$为335.4 eV，Pd $3d_{5/2}$为340.6 eV，峰间距为5.20 eV）相一致[35]，证明有钯原子的生成。摩尔比[NAC]/[Pd]=2∶1合成的Pd NPs，Pd $3d_{3/2}$和Pd $3d_{5/2}$的电子结合能分别为337.0 eV和342.3 eV，峰间距为5.30 eV，其电子结合能高于摩尔比[NAC]/[Pd]=1∶1合成的Pd NPs的电子结合能，主要是由于在合成Pd NPs的过程中生成了Pd^{II}-NAC的复合物，这与我们用UV-vis分析得出的结论相一致。

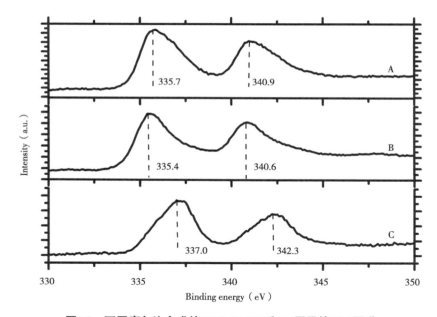

图2.3 不同摩尔比合成的NAC-Pd NPs和Pd原子的XPS图谱

（A为摩尔比[NAC]/[Pd]=1∶1，B为Pd原子，C为摩尔比[NAC]/[Pd]=2∶1）

2.3.4 NAC-Pd NPs的红外光谱

通过FTIR对NAC配体和钯原子的相互反应进行了研究，图2.4为NAC/Pd摩尔比为1∶1合成的NAC-Pd NPs（a）和NAC（b）的FTIR图。从文献[36]中可知NAC的FTIR，在3 200～3 500 cm^{-1}处的吸收归因于O—H的伸缩振动，应是来源于残留的水；在2 547 cm^{-1}处是NAC的特征峰，是由于S—H键伸缩振动；在1 718 cm^{-1}和1 535 cm^{-1}处的吸收分别归属于C=O的伸缩振动和N—H的弯曲振动。与之相比较，在NAC-Pd NPs的FTIR中，C=O的伸缩振动和N—H的弯曲振动的吸收分别移动于1 600 cm^{-1}和1 380 cm^{-1}处，说明钯原子会干扰NAC-Pd NPs的拉伸和振动吸收；而在2 547 cm^{-1}处的S—H键消失，证明钯原子与NAC配体中的—S基相互反应，生成了稳定的Pd—S键，而NAC作为配体包裹在钯原子表面，阻止了钯原子的聚集，配体NAC起到稳定和保护Pd NPs的作用，这与文献报道的一致[37-39]。

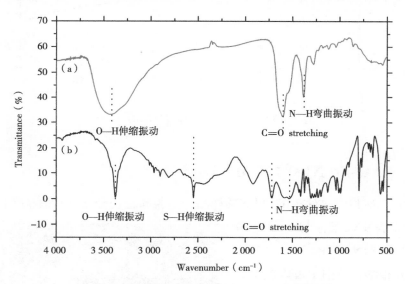

图2.4 NAC-Pd NPs（a）和NAC（b）的FTIR图

2.3.5 NAC-Pd NPs的TGA

TGA可以用来研究Pd NPs的热稳定性，并对其配体组分含量进行分析。在温度升到一定值时，Pd NPs的NAC巯基配体会脱离钯原子表面而发生热解吸附现象，造成Pd NPs的质量会产生变化。因此，可以通过NAC-Pd NPs质量发生的变化的温度与失去的重量，来考察Pd NPs的热稳定信息和其结合的NAC配体含量。图2.5为NAC配体与H_2PdCl_4摩尔比为1∶1的Pd NPs的TGA，如图所示，Pd NPs失重比例为22%，也就是说在NAC-Pd NPs中配体的含量约为22%。

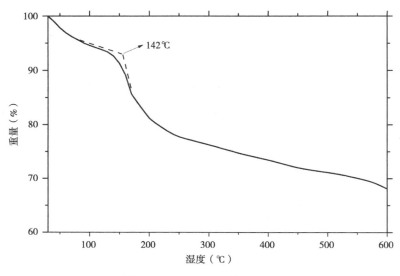

图2.5 NAC-Pd NPs的TGA图

2.3.6 NAC-Pd NPs的TEM和DLS

采用TEM和DLS对NAC-Pd NPs样品的形貌和粒径分布进行表征，图2.6和图2.7分别为NAC配体与H_2PdCl_4摩尔比为1∶1的NAC-Pd NPs的TEM和粒径分布图，从图中可以看出，NAC保护的Pd NPs近似球形，且分散性和粒径分布均一性好，我们选取100个Pd NPs测定其粒径大小，通过计算得出NAC-Pd NPs的平均粒径为（2.17±0.20）nm。图2.8显示了NAC配体与H_2PdCl_4摩尔比为1∶1的NAC-Pd NPs的流体动力学尺寸约为3.7 nm，比TEM的测量结果大，是由于水溶液中Pd NPs周围有溶剂层的缘故[40]。

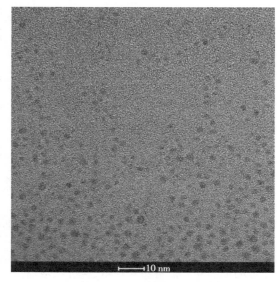

图2.6 NAC-Pd NPs的TEM图

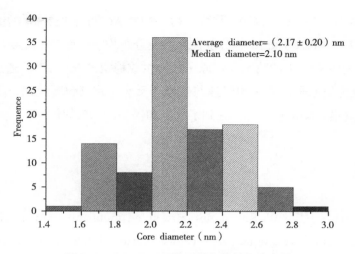

图2.7 NAC-Pd NPs的透射电镜粒径分布图

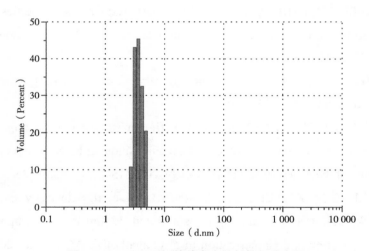

图2.8 NAC-Pd NPs的动态光散射粒径分布图

2.4 结论

本章采用NAC作为配体,在冰浴条件下,使用$NaBH_4$还原氯钯酸合成了NAC保护的水溶性Pd NPs。通UV-vis、XPS、FTIR、TGA和TEM对不同摩尔比合成的NAC-Pd NPs进行表征,得出NAC/Pd摩尔比大于2时,所合成的Pd NPs为Pd^{II}-

NAC的复合物；当NAC/Pd摩尔比为1∶1合成的NAC-Pd NPs为Pd^0的纳米粒子，粒径约为（2.17±0.20）nm，其分散性和粒径分布均一性好。由此得出，在此反应过程中，配体NAC中的巯基与钯原子结合生成稳定的Pd—S键，通过$NaBH_4$还原H_2PdCl_4成为钯原子，NAC也迅速与钯原子反应，将其包裹起来，阻止已生成Pd NPs的再聚集，NAC作为配体起到了稳定和保护Pd NPs的作用。

参考文献

[1] LU Y H, CHEN W. Sub-nanometre sized metal clusters: from synthetic challenges to the unique property discoveries[J]. Chemical society reviews, 2012, 41(9): 3594-3623.

[2] GHOSH C R, PARIA S. Core/shell nanoparticles: classes, properties, synthesis mechanisms, characterization, and applications[J]. Chemical reviews, 2012, 112(4): 2373-2433.

[3] GUIHEN E. Nanoparticles in modern separation science[J]. Trac-trends in analytical chemistry, 2013, 46: 1-14.

[4] BI Q Y, DU X L, LIU Y M, et al. Efficient subnanometric gold-catalyzed hydrogen generation via formic acid decomposition under ambient conditions[J]. Journal of the American chemical society, 2012, 134(21): 8926-8933.

[5] MORENO M, KISSELL L N, JASINSKI J B, et al. Selectivity and reactivity of alkylamine-and alkanethiolate-stabilized Pd and PdAg nanoparticles for hydrogenation and isomerization of allyl alcohol[J]. ACS Catal, 2012, 2(12): 2602-2613.

[6] KUMAR S S, KWAK K, LEE D. Electrochemical sensing using quantum-sized gold nanoparticles[J]. Analytical chemistry, 2011, 83(9): 3244-3247.

[7] MORENO M, IBAÑEZ F J, JASINSKI J B, et al. Hydrogen reactivity of palladium nanoparticles coated with mixed monolayers of alkyl thiols and alkyl amines for sensing and catalysis applications[J]. Journal of the American chemical society, 2011, 133(12): 4389-4397.

[8] TSENG R J, HUANG J X, OUYANG J Y, et al. Polyaniline nanofiber/gold

nanoparticle nonvolatile memory[J]. Nano letters, 2005, 5(6): 1077-1080.

[9] TAO A, KIM F, HESS C, et al. Langmuir-blodgett silver nanowire monolayers for molecular sensing using surface-enhanced raman spectroscopy[J]. Nano letters, 2003, 3(9): 1229-1233.

[10] YAVUZ M S, CHENG Y Y, CHEN J Y, et al. Gold nanocages covered by smart polymers for controlled release with near-infrared light[J]. Nat mater, 2009, 8(12): 935-939.

[11] XIAO L, ZHUANG L, LIU Y, et al. Activating Pd by morphology tailoring for oxygen reduction[J]. Journal of the American chemical society, 2009, 131(2): 602-608.

[12] COOKSON J. The preparation of palladium nanoparticles[J]. Platinum metals review, 2012, 56(2): 83-98.

[13] TAO A R, HABAS S, YANG P D. Shape control of colloidal metal nanocrystals[J]. Small, 2008, 4(3): 310-325.

[14] ZAMBORINI F P, GROSS S M, MURRAY R W. Synthesis, characterization, reactivity, and electrochemistry of palladium monolayer protected clusters[J]. Langmuir, 2001, 17(2): 481-488.

[15] LARIOS-RODRÍGUEZ E A, CASTILLÓN-BARRAZA F F, BORBÓN-GONZÁLEZ D J, et al. Green-chemical synthesis of monodisperse Au, Pd and bimetallic (core-shell) A-Pd, Pd-Au nanoparticles[J]. Advanced science, engineering and medicine, 2013, 5(7): 665-672.

[16] ROY P S, BAGCHI J, BHATTACHARYA S K. Synthesis of polymer-protected palladium nanoparticles of contrasting electrocatalytic activity: a comparative study with respect to reflux time and reducing agents[J]. Colloids and surfaces A: physicochemical and engineering aspects, 2010, 359(1): 45-52.

[17] BONET F, DELMAS V, GRUGEON S, et al. Synthesis of monodisperse Au, Pt, Pd, Ru and Ir nanoparticles in ethylene glycol[J]. Nanostructured materials, 1999, 11(8): 1277-1284.

[18] ALVAREZ G F, MAMLOUK M, SENTHIL KUMAR S M, et al. Preparation and characterisation of carbon-supported palladium nanoparticles for oxygen reduction in low temperature PEM fuel cells[J]. Journal of applied

electrochemistry, 2011, 41(8): 925-937.

[19] GIERSIG M, MULVANEY P. Preparation of ordered colloid monolayers by electrophoretic deposition[J]. Langmuir, 1993, 9(12): 3408-3413.

[20] SHARMA S, KIM B, LEE D. Water-Soluble Pd nanoparticles capped with glutathione: synthesis, characterization, and magnetic properties[J]. Langmuir, 2012, 28(45): 15958-15965.

[21] GAVIA D J, MAUNG M S, SHON Y S. Water-soluble Pd nanoparticles synthesized from ω-carboxyl-S-alkanethiosulfate ligand precursors as unimolecular micelle catalysts[J]. ACS Applied materials & interfaces, 2013, 5(23): 12432-12440.

[22] SAKAI N, TATSUMA T. One-step synthesis of glutathione-protected metal (Au, Ag, Cu, Pd, and Pt) cluster powders[J]. Journal of materials chemistry A, 2013, 1(19): 5915-5922.

[23] CARGNELLO M, WIEDER N L, CANTON P, et al. A versatile approach to the synthesis of functionalized thiol-protected palladium nanoparticles[J]. Chemistry of materials, 2011, 23(17): 3961-3969.

[24] XIA Y, YANG P, SUN Y, et al. One-dimensional nanostructures: synthesis, characterization, and applications[J]. Advanced materials, 2003, 15(5): 353-389.

[25] CHEN C, WANG L, JIANG G H, et al. Chemical preparation of special-shaped metal nanomaterials through encapsulation or inducement in soft solution[J]. 2006, 11(1): 34-40.

[26] HARADA T, IKEDA S, MIYAZAKI M, et al. A simple method for preparing highly active palladium catalysts loaded on various carbon supports for liquid-phase oxidation and hydrogenation reactions[J]. Journal of molecular catalysis A: chemical, 2007, 268(1-2): 59-64.

[27] NATH S, PRAHARAJ S, PANIGRAHI S, et al. Synthesis and characterization of N,N-dimethyldodecylamine-capped aucore-pdshell nanoparticles in toluene[J]. Langmuir, 2005, 21(23): 10405-10408.

[28] MUNK V P, SADLER P J. Palladium(ii) diamine complex induces reduction of glutathione disulfide[J]. Chemical communications, 2004(16): 1788-1789.

[29] VASIĆ V M, TOŠIĆ M S, NEDELJKOVIĆ J M. Influence of sodium dodecyl

sulphate micelles on the kinetics of complex formation between $Pd(H_2O)_4^{2+}$ and S-carboxymethyl-L-cysteine[J]. Journal of physical organic chemistry, 1996, 9(6): 398-402.

[30] FAKIH S, MUNK V A, SHIPMAN M A, et al. Novel adducts of the anticancer drug oxaliplatin with glutathione and redox reactions with glutathione disulfide[J]. European journal of inorganic chemistry, 2003, 2003(6): 1206-1214.

[31] YANG Z Q, SMETANA A B, SORENSEN C M, et al. Synthesis and characterization of a new tiara Pd(II) thiolate complex, $[Pd(SC_{12}H_{25})_2]_6$, and its solution-phase thermolysis to prepare nearly monodisperse palladium sulfide nanoparticles[J]. Inorganic chemistry, 2007, 46(7): 2427-2431.

[32] LI Z P, GAO J, XING X T, et al. Synthesis and characterization of n-alkylamine-stabilized palladium nanoparticles for electrochemical oxidation of methane[J]. Journal of physical chemistry C, 2009, 114(2): 723-733.

[33] HUANG X Q, TANG S H, MU X L, et al. Freestanding palladium nanosheets with plasmonic and catalytic properties[J]. Nat nanotechnol, 2011, 6(1): 28-32.

[34] VOOGT E H, MENS A J M, GIJZEMAN O L J, et al. XPS analysis of palladium oxide layers and particles[J]. Surface science, 1996, 350(1): 21-31.

[35] MOULDER J F, CHASTAIN J, KING R C. Handbook of x-ray photoelectron spectroscopy: a reference book of standard spectra for identification and interpretation of XPS data[J]. Physical electronics, 1995, 22(1): 7-10.

[36] BIERI M, BÜRGI T. Adsorption kinetics, orientation, and self-assembling of N-acetyl-l-cysteine on gold: a combined ATR-IR, PM-IRRAS, and QCM study[J]. The journal of physical chemistry B, 2005, 109(47): 22476-22485.

[37] SHIBU E S, MUHAMMED M A H, TSUKUDA T, et al. Ligand exchange of Au25SG18 leading to functionalized gold clusters: spectroscopy, kinetics, and luminescence[J]. The journal of physical chemistry C, 2008, 112(32): 12168-12176.

[38] FARRAG M. Preparation, characterization and photocatalytic activity of size selected platinum nanoclusters[J]. Journal of photochemistry and photobiology A: chemistry, 2016, 318: 42-50.

[39] FARRAG M, THÄMER M, TSCHURL M, et al. Preparation and spectroscopic properties of monolayer-protected silver nanoclusters[J]. The journal of physical chemistry C, 2012, 116(14): 8034-8043.

[40] SELIM K M K, KANG I K, GUO H Q. Albumin-conjugated cadmium sulfide nanoparticles and their interaction with KB cells[J]. Macromolecular research, 2009, 17(6): 403-410.

第3章

NAC保护的Pd NPs的色谱分离分析研究

第3章 NAC保护的Pd NPs的色谱分离分析研究

3.1 引言

由于纳米粒子通常是不同尺寸粒径纳米粒子组成的混合物，目前针对纳米粒子的分离方法有凝胶电泳法[1-2]、尺寸排阻色谱法[3-4]、离子交换色谱法[5]、分子印记法[6]、毛细管电泳法[7-8]、高速离心法[9]和高效液相色谱法[10]等。这些分离方法既有优点又有缺点，例如，由于纳米颗子的固定相的大表面积和高表面活性，尺寸排阻色谱法是采用色谱柱中填充物吸附纳米粒子达到分离的目的，但对不同样品的分离需要使用不同类型的色谱柱；虽然毛细管电泳法可以通过减少分离体系的表面效应来解决尺寸排阻色谱法存在的缺点[11]，但由于毛细管电泳法收集的分离样品组分量太少，大大限制了分离后纳米粒子组分的表征和应用；而利用高效液相色谱法来分离纳米粒子，一方面用单根色谱柱就可以达到分离的目的，另一方面又可以实现对分离样品组分的收集、分析和表征。Gong等[12]采用RP-HPLC分离碳纳米粒子，流动相为甲醇和pH值5.5醋酸缓冲液，结合梯度洗脱，根据紫外检测器色谱峰收集分离组分，共收集到13个具有代表的色谱峰，在室温下，采用氮气吹浓缩，将分离组分应用FL、uv-vis等分析，结果表明，所分离组分具有不同特征峰，证明所合成的碳纳米粒子一种复杂的混合物，通过高效液相色谱法分离得到的碳纳米粒子组分比与碳纳米粒子混合物有更优异的光学性质。Zhang等[13]利用荧光和紫外检测器的离子对高效液相色谱法成功地实现对金纳米粒子的分离，其采用C18色谱柱，研究了流动相中离子对试剂与甲醇含量对分离的影响以及分离组分的uv-vis和FL，结果表明，得到了12种金纳米分离组分，再结合所分离的金纳米组分的特征吸收光谱，确定分离组分的流出色谱柱的顺序是按照尺寸从小到大，并初步确定所合成的金纳米粒子中含有12～22个金原子。

TEM是研究纳米粒子尺寸的主要工具之一，但对于纳米粒子的确切尺寸特别是纳米粒径尺寸（<3 nm）仍然是一个巨大的挑战，是因为在成像过程中，由于电子束加热纳米粒子可能发生熔化效应。迄今为止，各种质谱技术，如激光解吸电离质谱（LDI-MS）[14-16]、基质辅助激光解析串联飞行时间质谱（MALDI-TOF MS）[10,17-19]、电喷雾质谱（ESI-MS）[20-22]、离子淌度质谱[23]、快原子轰击质谱[24]已被应用于研究纳米粒子的原子核心质量。到目前为止，MALDI-TOF MS已被证明是一种非常有效的方法来鉴定纳米粒子原子和配体的数量。Zhang等[25]将MALDI-TOF MS技术应用于分析L-组氨酸保护的金纳米粒子化学组成，结果表明，纳米粒子中含有10个、11个、12个、13个金原子结合不同数量L-组氨酸的混

合物。Arnold等[26]也将MALDI-TOF MS技术应用于硫醇基保护的金纳米粒子的化学组成，研究发现质谱峰之间的间距分子量相差约197 Da，相当于一个金原子。Whetten等[27]也使用MALDI-TOF MS分析金属纳米粒子，因此，MALDI-TOF MS是一种可靠的技术来测量小尺寸纳米粒子的分子量和化学组成。

因此，在文献的基础上，本章我们采用C18色谱柱（250 mm × 4.6 mm，5 μm），流动相为甲醇和$Bu_4N^+F^-$水溶液，结合梯度洗脱，利用反向离子对高效液相色谱法分离水溶性NAC-Pd NPs。研究了$Bu_4N^+F^-$和甲醇含量对NAC-Pd NPs分离的影响，利用UV-vis和MALDI-TOF MS对所分离的Pd NPs组成进行分析。

3.2 实验部分

3.2.1 主要试剂和仪器

$Bu_4N^+F^-$（>98%）购于国际实验室（San Bruno，CA，USA）；2,5-二羟基苯甲酸（DHB，98%）购于Sigma公司（St Louis，MO）；色谱级甲醇（MeOH）购于Bangkok公司（Thailand）；冰醋酸（CH_3COOH，北京化学试剂厂）；透析膜和醋酸纤维素管均购于北京索莱宝科技有限公司；醋酸铵（NH_4Ac，天津光复化学试剂有限公司）；氮气（N_2，99.99%，V/V，利泽工业气体经销，太原）；实验用水为超纯水（Mill-Q Advantage A10超纯水机，默克密理博）；其他所有试剂均为分析纯以上无须再纯化。

Free Zone 6真空冷冻干燥机（Labconco，美国）；Waters Alliance e2695高效液相色谱仪（Millford，MA，美国）包括分离系统，二极管阵列检测器（PDA）；Zorbax Eclipse Plus-C18色谱柱（250 mm × 4.6 mm，5 μm，Agilent Technologies，Santa Clara，CA，USA）；Bruker Autoflex基质辅助激光解吸电离-飞行时间质谱仪（MALDI-TOF-MS，Bremen，德国）；C18固相萃取柱（Alltech Associates Inc.）；固相萃取装置（天津博纳艾杰尔技术科技有限公司）；RE 52A真空旋转蒸发仪（上海亚荣生化仪器厂）。

3.2.2 NAC-Pd NPs的制备

Pd NPs合成具体步骤如下：称取0.442 5 g的$PdCl_2$溶于2 mol/L的HCl溶液中，

并定容至5 mL的容量瓶中,每次取0.5 mL其中含有0.25 mmol的Pd^{2+};称取2.04 g NAC溶于水中,并定容于20 mL的容量瓶中,按照NAC/Pd摩尔比为1∶1的要求取相应体积的NAC溶液。在冰浴条件下,分别用移液枪取0.3 mL的NAC溶液加入含有47.5 mL的甲醇的100 mL的圆底烧瓶中,逐滴加入0.5 mL的H_2PdCl_2溶液,强烈搅拌5 min后,再逐滴加入0.2 mol/L 12.5 mL的$NaBH_4$溶液,剧烈搅拌1 h后,旋转蒸发多余的甲醇溶液,将生成的沉淀物经透析膜透析3 d后,经真空冷冻干燥后,得到粉末状NAC/Pd摩尔比为1∶1的NAC-Pd NPs,将制备好的Pd NPs存于4℃冰箱中,备用。

3.2.3 高效液相色谱法的条件

色谱柱:安捷伦Zorbax Eclipse Plus-C18色谱柱(250 mm×4.6 mm,5 μm);检测器:PDA;检测波长:200~400 nm(PDA);进样量:20 μL;色谱柱温度:(22±1)℃;流速:0.70 mL/min;流动相:溶剂A为色谱级的甲醇,溶剂B为含有50 mmol/L $Bu_4N^+F^-$的溶液。流动相在使用前需用醋酸纤维素(0.45 μm)过滤膜过滤。

梯度洗脱程序如图3.1所示,0~15 min以50% V/V的流动相A洗脱,15~20 min将流动相A比例线性增加为55% V/V,20~30 min维持流动相A为55% V/V,30~35 min将流动相A比例线性增加为60% V/V,35~45 min维持流动相A为60% V/V,45~50 min将流动相A比例线性增加为65% V/V。其中Pd NPs经高效液相色谱系统分离后由PDA进行检测,使用紫外-可见信号收集分离的组分。

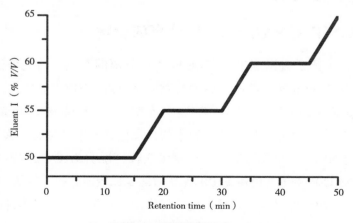

图3.1 梯度洗脱程序

3.2.4 NAC-Pd NPs的纯化

色谱分离收集的NAC保护的Pd NPs组分采用固相萃取法进行纯化。

（1）在室温下，用氮气将色谱分离收集的钯纳米溶液吹干，并用冰醋酸调pH值4后，备用。

（2）将C18固相萃取柱连接到固相萃取装置，依次用3 mL甲醇、3 mL水和3 mL pH值4的0.05 mol/L的醋酸铵缓冲溶液洗涤C18固相萃取柱。

（3）将步骤1中的钯纳米溶液上柱，再用pH值4，0.05 mol/L的醋酸铵缓冲溶液淋洗C18固相萃取柱，最后用甲醇洗脱柱子上的待测成分，在室温下将洗脱液用氮气吹干，从而得纯化的Pd NPs。

3.2.5 紫外-可见吸收光谱的测量

移取收集的NAC-Pd NPs分离组分溶液至比色皿中，空白为水溶液，用Varian Cary300紫外-可见分光光度计扫描200～500 nm内的吸收光谱。

3.2.6 质谱的测定

移取收集的NAC-Pd NPs分离组分甲醇溶液与1.0 mol/L的DHB溶液以1∶1 V/V 混合，取混合液4.0 μL滴于质谱样品板上，晾干后，用于MALDI-TOF-MS质谱仪进行测定。

3.3 结果与讨论

3.3.1 $Bu_4N^+F^-$对NAC-Pd NPs分离的影响

本章研究了离子对试剂$Bu_4N^+F^-$对Pd NPs分离的影响，首先考察了流动相不添加离子对试剂$Bu_4N^+F^-$的Pd NPs的高效液相色谱（HPLC）分离。如图3.2所示，NAC-Pd NPs在甲醇/水（1/1，V/V）为流动相中，检测波长为250 nm，色谱柱为C18硅胶柱的分离色谱图，并与纯水的分离色谱图做了对比。从图中可以看出，NAC-Pd NPs的色谱图只有一个色谱吸收峰，说明NAC-Pd NPs并没有被分离。另外，NAC-Pd NPs在色谱图中的保留时间也比较短，很快就从C18硅胶柱中流出，说明Pd NPs并没有与C18柱发生任何作用，只是从色谱柱内部的空隙之间直接流出，从而使Pd NPs在C18硅胶色谱柱中的保留时间很短。

第 3 章　NAC 保护的 Pd NPs 的色谱分离分析研究

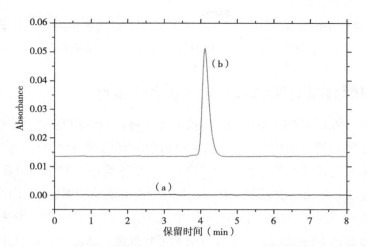

图3.2　NAC-Pd NPs在甲醇/水（1∶1 V/V）流动相中的高效液相色谱分离图

我们在流动相中加入了离子对试剂$Bu_4N^+F^-$，可以使NAC-Pd NPs和C18固定相发生作用，从而取得更好的分离效果。图3.3为C18色谱柱、离子对试剂和NAC-Pd NPs的作用示意图。由第2章研究可知，NAC配体通过Pd—S键与钯原子结合，NAC包裹其外面，从而合成NAC-Pd NPs，因此，在Pd NPs的表面存在NAC上的羧基（—COO^-），其可以和Bu_4N^+形成离子对[28]，再与色谱柱发生相互作用，从而达到理想的分离效果。如图3.3所示，有许多的NAC-Pd NPs分离峰被分离出来，表明在流动相中加入离子对试剂可以Pd NPs得到很好的分离，虽然Pd NPs的NAC配体都是一致的，但是，不同Pd NPs粒径同离子对试剂的作用可以决定钯纳

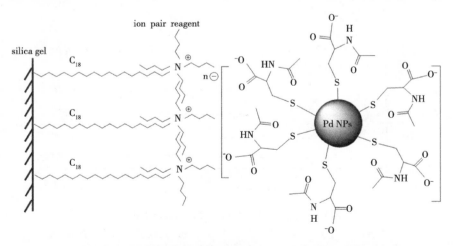

图3.3　色谱柱、二丁基氟化铵和Pd NPs的作用示意图

米在C18色谱柱中的保留时间。流动相中添加离子对试剂对具有强非极性固定相的Pd NPs样品的色谱分离的影响与离子对RP-HPLC分离理论相一致[13,29]。

3.3.2 甲醇含量对NAC-Pd NPs分离的影响

多分散的纳米粒子RP-HPLC分离依赖于非极性固定相和金属纳米粒子的封闭端配体之间的相互作用[10,12,30]。在流动相中的有机溶剂含量可以调节疏水性固定相在液相色谱柱中的保留时间[19,29,31]，从而影响纳米粒子的分离效果。本章研究了流动相中甲醇含量对NAC-Pd NPs分离的影响。图3.4为不同甲醇含量（40%～65%，V/V）对NAC-Pd NPs（3.0 mg/mL）样品RP-HPLC分离的影响，流动相为50 mmol/L的$Bu_4N^+F^-$溶液与不同甲醇含量。从图中可以看出，当甲醇含量为65%（V/V）时，保留时间在10 min左右出现了一个很大的洗脱峰，这应该是几个NAC-Pd NPs色谱峰被同时洗脱出来并聚在一起。当甲醇含量逐渐减少时，可以观察到很好的Pd NPs色谱分离峰，并且有更的多钯纳米色谱峰在随后的时间被洗脱出来，表明Pd NPs样品中含有不同尺寸粒径的NAC-Pd NPs组分。另外，当甲醇含量太少时（≤40%，V/V），后面的色谱分离峰有可能被保留在色谱柱上，不能被完全洗脱出来。研究表明，保留时间短的NAC-Pd NPs在甲醇含

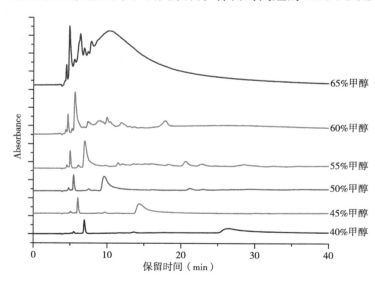

图3.4 甲醇含量对NAC-Pd NPs分离的影响

流动相A为甲醇溶液，流动相B为50 mmol/L $Bu_4N^+F^-$；
色谱图均为PDA检测器250 nm处的紫外-可见吸收色谱图

量较低时有较好的色谱分离效果，当甲醇含量逐渐增加时，保留时间长的NAC-Pd NPs可以被洗脱出来，但保留时间短的NAC-Pd NPs又容易聚集在一起，因此，我们在随后的实验中采用流动相中甲醇含量由以低至高逐渐增加的梯度洗脱程序以完全洗脱样品中的NAC-Pd NPs组分。

3.3.3 梯度洗脱对NAC-Pd NPs的分离

为了达到更好的分离效果，我们将梯度洗脱程序应用于多分散性的NAC-Pd NPs（3.0 mg/mL）的RP-HPLC的分离。如图3.5所示，A为NAC-Pd NPs和B纯水采用梯度洗脱程序的紫外-可见吸收色谱图，PDA检测器波长为250 nm。作为纯水对照的色谱分离图B，基线随着洗脱时间的增加发生漂移，是由于梯度洗脱程序中流动相甲醇含量的增加而造成的。NAC-Pd NPs的色谱分离图A在250 nm波长下，根据紫外色谱峰中至少有11个钯纳米组分被分离出来，在50 min后没有发现洗脱峰。所用标注的分离组分被收集后，采用固相萃取法进行纯化，用于紫外-可见光谱和质谱（MALDI-TOF-MS）的分析。其余未被标识的组分，由于含量较低，无法用于质谱的分析，因此并没有体现在我们的研究内容中。

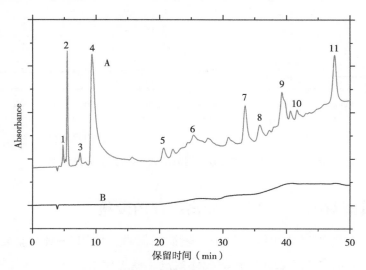

图3.5 NAC-PdNPs样品的色谱分离图

图3.6为图3.5中被标注的11个NAC-Pd NPs分离组分的UV-vis。从图中可以看出，所有分离组分色谱峰的UV-vis显示出比合成的NAC-Pd NPs产物更加明显的

光谱特征[32-33]。为了便于比较，组分1和组分7扩大了2倍，组分3扩大了10倍。大多组分在300 nm以下都表现出较强的吸收，并且在较长的波长（>400 nm）处没有发现明显的吸收带。组分1～3，组分7和组分11表现出更独特的特征吸收峰。组分1和组分2在240 nm处有吸收峰，组分3和组分7～11分别在230 nm和260 nm处有清晰的吸收峰，这些组分特征峰隐藏于NAC-Pd NPs样品混合物。实际上，组分紫外光谱吸收峰影响合成产物Pd NPs紫外光谱中200～400 nm处的宽吸收带，见第2章图2.2 NAC-Pd NPs的紫外-可见吸收光谱图。所有分离组分的吸收峰都没有特有的SPR，表明分离的NAC-Pd NPs的粒径都小于3 nm[15,34-35]。研究结果表明，色谱分离和UV-vis结合是评估钯纳米分离组分特征光谱的有效办法，NAC/Pd摩尔比为1∶1合成的NAC-Pd NPs是一种混合物，其紫外-可见吸收光谱仅代表所有单独分离的Pd NPs的总和。

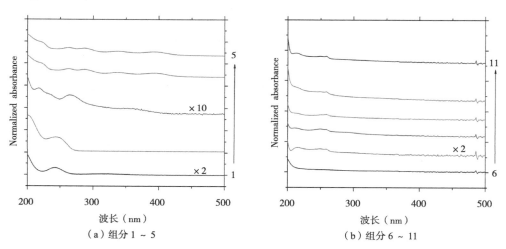

图3.6 NAC-Pd NPs分离组分的紫外-可见吸收光谱图

3.3.4 质谱对NAC-Pd NPs分离组分的分析

TEM被广泛用来金属纳米粒子的表征，但是，使用TEM不能准备确定金属纳米粒子的钯原子和表面保护的配体数量。通过质谱方法可以提供TEM所不能提供的信息[17,27,36]，此外，质谱技术比只能表征纳米粒子表面配体的NMR时间更快和需要更少的样品量。图3.7为使用MALDI-TOF MS对NAC-Pd NPs的分离组分1-11分析的质谱图。图中用$Pd_x(NAC)_y$的形式来表示，其中x代表分离组分中钯原子数，y代表分离组分中NAC的配体数。表3.1为质谱峰$Pd_x(NAC)_y$的

第3章 NAC保护的Pd NPs的色谱分离分析研究

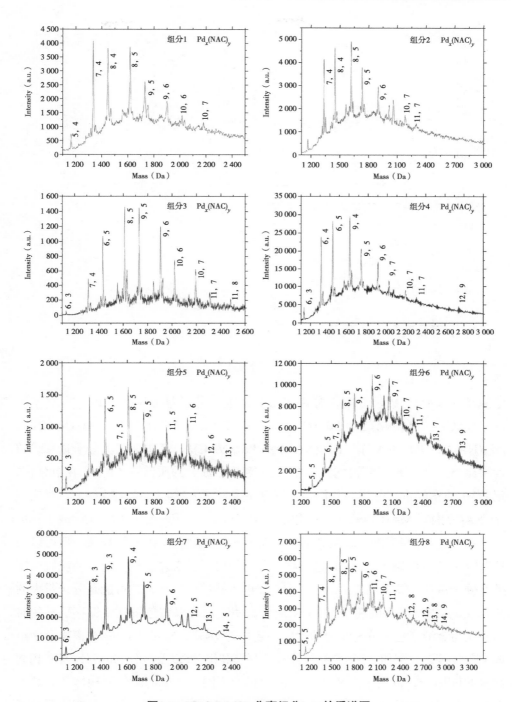

图3.7 NAC-Pd NPs分离组分1-11的质谱图

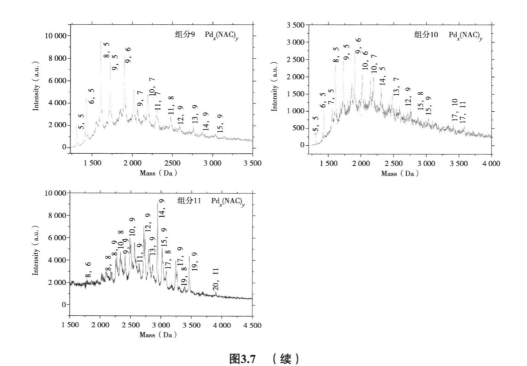

图3.7 （续）

钯原子和NAC的数目分配。分离组分1-11最大分子量分别代表$Pd_{10}(NAC)_7$、$Pd_{11}(NAC)_7$、$Pd_{11}(NAC)_8$、$Pd_{12}(NAC)_9$、$Pd_{13}(NAC)_6$、$Pd_{13}(NAC)_9$、$Pd_{14}(NAC)_5$、$Pd_{14}(NAC)_9$、$Pd_{15}(NAC)_9$、$Pd_{17}(NAC)_{11}$和$Pd_{20}(NAC)_{11}$。其中，最大分子量基线的信噪比≥3。图3.8为组分1的质谱图（a）和矩形放大的质谱图（b），在组分1选定的3个质谱峰进行分析，它们之间有分子量差值为284（±1~2 m/z），对应着1个NAC配体（M=163 m/z），Pd原子（M=106 m/z）和—CH_3（M=15 m/z），可见，每个NAC-Pd NPs分离组分都具有其特有分子量和NAC配体数量，但可以通过RP-HPLC方法确定其在C18色谱柱上的保留时间。总之，分离组分的洗脱顺序遵循分子量从小到大的先后，依次从C18色谱柱洗脱出来，这可能是分子量大的NAC-Pd NPs与更多的离子对Bu_4N^+结合，使它们可以在色谱柱中保存更长的时间，这与金纳米色谱分离组分的现象一致[8,37]。尽管，NAC-Pd NPs分离组分中组成中非常接近，例如，组分2和组分3，组分5和组分6以及组分7和组分8具有相同数量的Pd原子，但由于在NAC配体的数量的不同，形成的配位异构体[38]。总的来说，我们的结果进一步证实，合成的NAC-Pd NPs样品实际上是具有不同分子量的各种类型Pd NPs的复杂混合物。

第 3 章 NAC 保护的 Pd NPs 的色谱分离分析研究

表 3.1 质谱峰 $Pd_x(NAC)_y$ 的分子量

分子式	m/z（Da）	分子式	m/z（Da）
$Pd_5(NAC)_4$	1 164	$Pd_{11}(NAC)_6$	2 149
$Pd_5(NAC)_5$	1 327	$Pd_{11}(NAC)_7$	2 312
$Pd_6(NAC)_3$	1 128	$Pd_{11}(NAC)_8$	2 475
$Pd_6(NAC)_5$	1 454	$Pd_{11}(NAC)_9$	2 638
$Pd_7(NAC)_4$	1 397	$Pd_{12}(NAC)_8$	2 581
$Pd_7(NAC)_5$	1 560	$Pd_{12}(NAC)_9$	2 744
$Pd_8(NAC)_3$	1 341	$Pd_{13}(NAC)_5$	2 199
$Pd_8(NAC)_4$	1 504	$Pd_{13}(NAC)_6$	2 362
$Pd_8(NAC)_5$	1 667	$Pd_{13}(NAC)_7$	2 525
$Pd_8(NAC)_6$	1 830	$Pd_{13}(NAC)_8$	2 688
$Pd_8(NAC)_8$	2 156	$Pd_{13}(NAC)_9$	2 851
$Pd_9(NAC)_3$	1 447	$Pd_{14}(NAC)_5$	2 305
$Pd_9(NAC)_4$	1 610	$Pd_{14}(NAC)_9$	2 957
$Pd_9(NAC)_5$	1 773	$Pd1_{15}(NAC)_8$	2 901
$Pd_9(NAC)_6$	1 936	$Pd_{15}(NAC)_9$	3 064
$Pd_9(NAC)_7$	2 099	$Pd_{17}(NAC)_9$	3 276
$Pd_9(NAC)_8$	2 262	$Pd_{17}(NAC)_{10}$	3 439
$Pd_9(NAC)_9$	2 425	$Pd_{17}(NAC)_{11}$	3 602
$Pd_{10}(NAC)_6$	2 042	$Pd_{19}(NAC)_8$	3 326
$Pd_{10}(NAC)_7$	2 205	$Pd_{19}(NAC)_9$	3 489
$Pd_{11}(NAC)_5$	1 986	$Pd_{20}(NAC)_{11}$	3 922

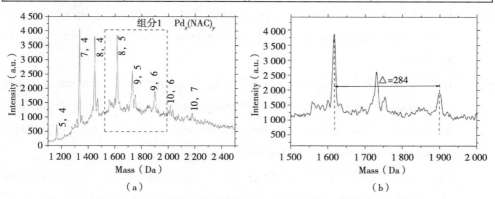

图 3.8 组分 1 的质谱图（a）和矩形放大的质谱图（b）

3.4 结论

本章利用RP-HPLC对NAC-Pd NPs进行分离，得到11个钯纳米的分离组分，并对其进行UV-vis和质谱分析，表明每一个分离组分都有特征的UV-vis，这些性质是通过研究NAC-Pd NPs混合物无法得到的。通过研究Pd NPs分离组分的MALDI-TOF-MS，得到每一个分离组分的化学组成，分别为$Pd_{10}(NAC)_7$、$Pd_{11}(NAC)_7$、$Pd_{11}(NAC)_8$、$Pd_{12}(NAC)_9$、$Pd_{13}(NAC)_6$、$Pd_{13}(NAC)_9$、$Pd_{14}(NAC)_5$、$Pd_{14}(NAC)_9$、$Pd_{15}(NAC)_9$、$Pd_{17}(NAC)_{11}$和$Pd_{20}(NAC)_{11}$。本章的研究为纳米材料范围内的Pd NPs及其他贵金属纳米粒子的分离、性质研究提供了研究基础。

参考文献

[1] HANAUER M, PIERRAT S, ZINS I, et al. Separation of nanoparticles by gel electrophoresis according to size and shape[J]. Nano letters, 2007, 7(9): 2881-2885.

[2] SOKOŁOWSKA K, MALOLA S, LAHTINEN M, et al. Towards controlled synthesis of water-soluble gold nanoclusters: synthesis and analysis[J]. The journal of physical chemistry C, 2019, 123(4): 2602-2612.

[3] TSUNOYAMA H, NEGISHI Y, TSUKUDA T. Chromatographic isolation of "missing" Au_{55} clusters protected by alkanethiolates[J]. Journal of the American chemical society, 2006, 128(18): 6036-6037.

[4] PITKÄNEN L, STRIEGEL A M. Size-exclusion chromatography of metal nanoparticles and quantum dots[J]. TrAC Trends in analytical chemistry, 2016, 80: 311-320.

[5] BOS W, STEGGERDA J J, YAN S P, et al. Separation of cationic metal cluster compounds by reversed phase HPLC[J]. Inorganic chemistry, 1988, 27(5): 948-951.

[6] KOENIG S, CHECHIK V. Au nanoparticle-imprinted polymers[J]. Chemical communications, 2005(32): 4110-4112.

[7] LO C K, PAAU M C, XIAO D, et al. Application of capillary zone electrophoresis for separation of water-soluble gold monolayer-protected

clusters[J]. Electrophoresis, 2008, 29(11): 2330-2339.

[8] PAAU M C, LO C K, YANG X P, et al. Capillary electrophoretic study of thiolated α-cyclodextrin-capped gold nanoparticles with tetraalkylammonium ions[J]. Journal of chromatography A, 2009, 1216(48): 8557-8562.

[9] ZOOK J M, RASTOGI V, MACCUSPIE R I, et al. Measuring agglomerate size distribution and dependence of localized surface plasmon resonance absorbance on gold nanoparticle agglomerate size using analytical ultracentrifugation[J]. ACS Nano, 2011, 5(10): 8070-8079.

[10] XIE S P, PAAU M C, ZHANG Y, et al. High-performance liquid chromatographic analysis of as-synthesised N,N′-dimethylformamide-stabilised gold nanoclusters product[J]. Nanoscale, 2012, 4(17): 5325-5332.

[11] SCHNABEL U, FISCHER C H, KENNDLER E. Characterization of colloidal gold nanoparticles according to size by capillary zone electrophoresis[J]. Journal of microcolumn, 1997, 9: 529.

[12] GONG X J, HU Q, CHIN P M, et al. High-performance liquid chromatographic and mass spectrometric analysis of fluorescent carbon nanodots[J]. Talanta, 2014, 129: 529-538.

[13] ZHANG Y, SHUANG S M, DONG C, et al. Application of HPLC and MALDI-TOF MS for studying as-synthesized ligand-protected gold nanoclusters products[J]. Analytical chemistry, 2009, 81(4): 1676-1685.

[14] SCHAAFF T G, WHETTEN R L. Controlled etching of Au:SR cluster compounds[J]. The journal of physical chemistry B, 1999, 103(44): 9394-9396.

[15] JIMENEZ V L, GEORGANOPOULOU D G, WHITE R J, et al. Hexanethiolate monolayer protected 38 gold atom cluster[J]. Langmuir, 2004, 20(16): 6864-6870.

[16] YAN B, ZHU Z J, MIRANDA O R, et al. Laser desorption/ionization mass spectrometry analysis of monolayer-protected gold nanoparticles[J]. Analytical bioanalytical chemistry, 2010, 396(3): 1025-1035.

[17] YANG X P, SU Y, PAAU M C, et al. Mass spectrometric identification of water-soluble gold nanocluster fractions from sequential size-selective precipitation[J].

Analytical chemistry, 2012, 84(3): 1765-1771.

[18] HU Q, PAAU M C, ZHANG Y, et al. Green synthesis of fluorescent nitrogen/sulfur-doped carbon dots and investigation of their properties by HPLC coupled with mass spectrometry[J]. RSC Advances, 2014, 4(35): 18065-18073.

[19] HU Q, PAAU M C, CHOI MARTIN M F, et al. Better understanding of carbon nanoparticles via high-performance liquid chromatography-fluorescence detection and mass spectrometry[J]. Electrophoresis, 2014, 35(17): 2454-2462.

[20] NEGISHI Y C, TAKASUGI Y, SATO S, et al. Kinetic stabilization of growing gold clusters by passivation with thiolates[J]. The journal of physical chemistry B, 2006, 110(25): 12218-12221.

[21] FIELDS-ZINNA C A, CROWE M C, DASS A, et al. Mass spectrometry of small bimetal monolayer-protected clusters[J]. Langmuir, 2009, 25(13): 7704-7710.

[22] FIELDS-ZINNA C A, PARKER J F, MURRAY R W. Mass spectrometry of ligand exchange chelation of the nanoparticle$[Au_{25}(SCH_2CH_2C_6H_5)_{18}]^{1-}$ by $CH_3C_6H_3(SH)_2$[J]. Journal of the American chemical society, 2010, 132(48): 17193-17198.

[23] ANGEL L A, MAJORS L T, DHARMARATNE A C, et al. Ion mobility mass spectrometry of $Au_{25}(SCH_2CH_2Ph)_{18}$ nanoclusters[J]. ACS Nano, 2010, 4(8): 4691-4700.

[24] DASS A, DUBAY G R, FIELDS-ZINNA C A, et al. FAB mass spectrometry of $Au_{25}(SR)18$ nanoparticles[J]. Analytical chemistry, 2008, 80(18): 6845-6849.

[25] ZHANG Y, HU Q, PAAU M C, et al. Probing histidine-stabilized gold nanoclusters product by high-performance liquid chromatography and mass spectrometry[J]. The journal of physical chemistry C, 2013, 117(36): 18697-18708.

[26] ARNOLD R J, REILLY J P. High-resolution time-of-flight mass spectra of alkanethiolate-coated gold nanocrystals[J]. Journal of the American chemical society, 1998, 120(7): 1528-1532.

[27] WHETTEN R L, KHOURY J T, ALVAREZ M M, et al. Nanocrystal gold molecules[J]. Advanced materials, 1996, 8(5): 428-433.

[28] YAO H, FUKUI T, KIMURA K. Chiroptical responses of d-/l-penicillamine-capped gold clusters under perturbations of temperature change and phase transfer[J]. The journal of physical chemistry C, 2007, 111(41): 14968-14976.

[29] CHOI M M F, DOUGLAS A D, MURRAY R W. Ion-pair chromatographic separation of water-soluble gold monolayer-protected clusters[J]. Analytical chemistry, 2006, 78(8): 2779-2785.

[30] ZHANG L, LI Z P, ZHANG Y, et al. High-performance liquid chromatography coupled with mass spectrometry for analysis of ultrasmall palladium nanoparticles[J]. Talanta, 2015, 131: 632-639.

[31] STÅHLBERG J. Retention models for ions in chromatography1[J]. Journal of chromatography A, 1999, 855(1): 3-55.

[32] GAUTIER C, BÜRGI T. Chiral N-isobutyryl-cysteine protected gold nanoparticles: preparation, size selection, and optical activity in the UV-vis and infrared[J]. Journal of the American chemical society, 2006, 128(34): 11079-11087.

[33] SAKAI N, TATSUMA T. One-step synthesis of glutathione-protected metal (Au, Ag, Cu, Pd, and Pt) cluster powders[J]. Journal of materials chemistry A, 2013, 1(19): 5915-5922.

[34] XIONG Y, XIA Y. Shape-controlled synthesis of metal nanostructures: the case of palladium[J]. Advanced materials, 2007, 19(20): 3385-3391.

[35] DANIEL M C, ASTRUC D. Gold nanoparticles: assembly, supramolecular chemistry, quantum-size-related properties, and applications toward biology, catalysis, and nanotechnology[J]. Chemical reviews, 2004, 104(1): 293-346.

[36] KOUCHI H, KAWASAKI H, ARAKAWA R. A new matrix of MALDI-TOF MS for the analysis of thiolate-protected gold clusters[J]. Analytical methods, 2012, 4(11): 3600-3603.

[37] FAKIH S, MUNK V P, SHIPMAN M A, et al. Novel adducts of the anticancer drug oxaliplatin with glutathione and redox reactions with glutathione disulfide[J]. European journal of inorganic chemistry, 2003(6): 1206-1214.

[38] NIIHORI Y, MATSUZAKI M, PRADEEP T, et al. Separation of precise compositions of noble metal clusters protected with mixed ligands[J]. Journal of the American chemical society, 2013, 135(13): 4946-4949.

第4章

DMF保护的Pd NPs的制备和表征

4.1 引言

　　贵金属纳米粒子由于特有的性质，使其在电化学[1]、新材料[2]、催化化学[3]、环境化学[4]、物理学[5]等众多领域具有可观的应用前景，因此，制备方法的完善、新应用工艺的开发也逐渐成为科研工作者关注的重点。在合成反应过程中，贵金属前驱体、还原剂、配体的合理选择可以制备出各种贵金属纳米粒子，通过反应物配比的调整，配体的官能化可用来控制合成纳米粒子的粒径、稳定性、溶解性等性质。

　　目前，化学法合成Pd NPs的主要有热分解法[6-9]、超声波辐射[10-12]、微波辐射[13-14]、光化学[15-17]和化学还原法[18-24]等，其中化学还原法是比较成熟的合成方法之一。同时，制备Pd NPs时常用的有机配体有巯基化合物[25]、表面活性剂[26]、聚合物[27]和枝状化合物[28]等。DMF是一种常见的化学试剂，在工业上有广泛的用途，可以用作化工原料，另外，DMF也是优良的有机溶剂，可作为高聚物聚乙烯、聚丙烯腈和聚酰胺的溶剂，也可用作萃取剂、医药工业生产激素和农药杀虫脒的原料[29]，因此，DMF作为有机配体保护重金属纳米粒子也被科研人员关注。例如，2008年，Liu等[30]利用DMF作为溶剂和还原剂，研发了一种制备高荧光性、分散性好的金纳米粒子的新途径。首先，在室温下，将$HAuCl_4$的水溶液与DMF水溶液混合，然后在剧烈搅拌下，加热至140℃下回流4 h，而后冷却至室温后，离心，在真空下蒸发掉过量溶剂后，将残余物通过甲醇为洗脱剂的硅胶柱，最后，将收集到的金纳米粒子，通过UV-vis、FL和TEM表征，结果表明，获得了具有高荧光特性的DMF为配体保护的金纳米粒子。Kawasaki等[31]通过改进Liu的合成方法，也合成了DMF保护的金纳米粒子，研究表明该纳米粒子可以分别在甲醇溶液和不同pH的水溶液中稳定地保持1个月且性质不会发生变化。Hideya等[32]合成了DMF保护的铂纳米粒子，研究表明，DMF-Pt NPs具有荧光特性，可以在pH值2、pH值12的水溶液和1 mol/L的氯化钠水溶液中保持高度的分散性。

　　本文采用一步法合成DMF作为配体保护的水溶性Pd NPs（DMF-Pd NPs），其中DMF起着溶液、配体和还原剂三重作用。通过UV-vis、FL、FITR、TGA、TEM和DLS对合成的DMF-Pd NPs进行表征。

4.2 实验部分

4.2.1 主要试剂和仪器

六氯化钯酸铵[$(NH_4)_2PdCl_6$，M=355.22 g/mol，>99.9%]购于Sigma-Aldrich试剂公司；DMF（>99%）购于Bangkok公司（Thailand）；氮气（N_2，99.99%，V/V，利泽工业气体经销，太原）；实验用水为超纯水（Mill-Q Advantage A10超纯水机，默克密理博）；其他所有试剂均为分析纯以上。

RE 52A真空旋转蒸发仪（上海亚荣生化仪器厂）；ME 204万分之一天平（METLER TLIEDO，上海）；Varian Cary300紫外-可见分光光度计（Varian，Palo Alto，CA）；QM4荧光光度计（Birmingham，NJ，美国）；JEOL JEM-1011透射电子显微镜（Tokyo，日本电子光学公司）；Perkin-Elmer Paragon 100 FTIR红外光谱仪（Birmingham，NJ）；Perkin-Elmer TGA 6热重分析仪（Waltham，MA）；Nano-ZS90 Zetasizer动态光散射分析仪（英国马尔文公司）；DF-101S集热式恒温加热磁力搅拌器（巩义市予华仪器有限责任公司）。

4.2.2 DMF-Pd NPs的制备

将150 μL 0.1 mol/L的$(NH_4)_2PdCl_6$水溶液加入到15 mL 140℃的DMF溶液中，强烈搅拌不同的时间（0.5 h、1 h、2 h、3 h、4 h、5 h、6 h、8 h），在80℃下，真空旋转（<10 mg Hg）蒸发多余的溶剂，在室温下，经氮气吹干，得到不同反应时间DMF保护的Pd NPs（DMF-Pd NPs），放干燥器中，备用。

4.2.3 UV-vis的测量

移取一定浓度的DMF-Pd NPs水溶液至比色皿中，空白为水溶液，用Varian Cary 300紫外-可见分光光度计扫描200～500 nm内的吸收光谱。

4.2.4 FL的测量

DMF-钯纳米纳米粒子溶液分别置于荧光杯中，在荧光仪上扫描FL，激发和发射狭缝设为10 nm，实验温度为（22±1）℃。

4.2.5 FTIR的表征

取DMF-Pd NPs样品并加入干燥的KBr粉末，在红外干燥灯下研磨均匀，然后通过压力控制机制成压片，测试波长为4 000~500 cm^{-1}，通过Perkin-Elmer Paragon 100 FTIR红外光谱仪进行红外表征。

4.2.6 TEM的表征

用TEM对合成的DMF-Pd NPs的分散情况和表面状态进行观察，将浓缩后的DMF-Pd NPs甲醇溶液，超声分散20 min，用移液枪将其滴至涂覆了碳膜的铜网上（300目），待溶剂自然挥发后，得到样品。TEM测定的操作电压为200 kV，通过对TEM中的纳米粒子进行测量统计，得到DMF-Pd NPs的大小尺寸及粒径分布。

4.2.7 TGA的测定

TGA在Perkin-Elmer TGA 6热重分析仪上测定，在氮气氛围下，称取一定量的DMF-Pd NPs样品，置于热重分析仪的坩埚内，以10 ℃/min升温速度，在20~800 ℃下，记录其失重比例和温度变化。

4.3 结果与讨论

4.3.1 DMF-Pd NPs的合成

图4.1为DMF-Pd NPs的合成示意图。在此Pd NPs合成反应中DMF同时起着溶剂、配体和还原剂的三重作用，首先，当DMF在加热温度高于100 ℃后会释放出CO和二甲胺，CO作为钯离子还原为零价钯原子[33]；其次，DMF作为配体和钯离子配位形成配合物，包裹在钯纳米外围，防止Pd NPs的聚集，从而形成DMF-Pd

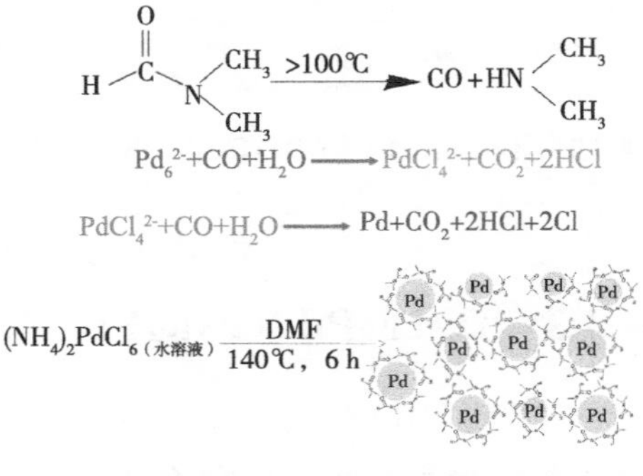

图4.1 DMF-Pd NPs合成示意图

NPs[30,34]。采用此方法合成的DMF-Pd NPs，易溶于水和甲醇等有机试剂。

4.3.2 DMF-Pd NPs的UV-vis

UV-vis常被用来考察Pd NPs光学性质的表征手段技术之一。在紫外-可见光区域存在吸收带是贵金属纳米粒子特性之一，研究表明，吸收带的出现主要是由贵金属粒子的表面等离子共振带（SPR）激发所引起，所以当贵金属纳米粒子的粒径大小及其配体不同时，其SPR必然会发生相应的变化[4,35-37]。

图4.2为不同反应时间的DMF-Pd NPs的UV-vis。在合成过程中，分别取不同的反应时间内的DMF-Pd NPs，经稀释一定倍数后测出不同反应时间内的紫外-可见吸收光谱图，Pd^{4+}在290 nm附近处吸收峰，随着反应时间的增加也逐渐消失，DMF-Pd NPs的吸光度随波长的增长而逐渐减小。图4.3所示，$(NH_4)_2PdCl_6$和DMF-Pd NPs溶液（反应时间6 h）的UV-vis，其中$(NH_4)_2PdCl_6$溶液在217 nm有1个较小的吸收峰，而化学合成得到的DMF-Pd NPs溶液的UV-vis有明显的光谱吸收峰，在250～400 nm，有一个宽的吸收带，是钯原子表面与DMF结合[38]，另外，从DMF-Pd NPs的UV-vis没有UV-visSPR的出现[4]，证明生成了DMF保护的Pd NPs。

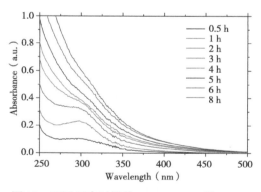 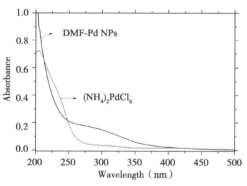

图4.2　不同反应时间的DMF-Pd NPs的UV-vis　　图4.3　氯钯酸铵和DMF-Pd NPs的UV-vis

4.3.3 DMF-Pd NPs的FL

一般采用FL研究贵金属纳米材料的光致发光特性，有较高的灵敏度，相关纳米粒子发光行为的研究报道也很多[39-44]。目前，关于钯纳米材料的FL报道很少，我们通过对DMF-Pd NPs荧光发光行为的研究，为其他Pd NPs材料的研究和应用提供一定的理论基础。图4.4为不同反应时间的DMF-Pd NPs溶液的FL。在合成过程中，分别取不同的反应时间内的DMF-Pd NPs，经稀释相同倍数后，激发波长固定为322 nm，在370～600 nm波长，测出不同反应时间的荧光谱图。由图可以得出，随着合成反应时间的增加，相同浓度的不同反应时间的Pd NPs的荧光强度也逐渐增强，当加热到6 h后，荧光强度变化幅度不大，基本保持不变。图4.5不同激发波长下DMF-Pd NPs溶液的荧光光谱。我们选用反应时间为6 h合成的DMF-Pd NPs，经稀释后，采用不同激发波长，分别为300 nm、322 nm、400 nm、450 nm和500 nm，在300～700 nm波长，测出不同激发波长的荧光谱图。从图中可以看出，随着激发波长的变化，发射波长位置发生红移，Pd NPs的荧光发射波长的可能是电子从$4d^{10}$带到$5sp$带的跃迁，与金纳米的荧光发光特性相类似[31]。

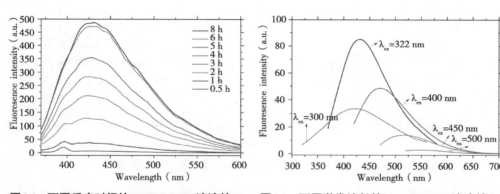

图4.4　不同反应时间的DMF-Pd NPs溶液的FL　　图4.5　不同激发波长的DMF-Pd NPs溶液的FL

4.3.4　DMF-Pd NPs的FTIR

取反应时间为6 h合成的DMF-Pd NPs样品并加入干燥的KBr粉末，在红外干燥灯下研磨均匀，通过压力控制机压制成压片，扫描波长为4 000～500 cm^{-1}，通过Perkin-Elmer Paragon 100 FTIR红外光谱仪进行红外表征，测试结果如图4.6。图中A曲线为DMF的FTIR，DMF FTIR的从文献中可以得知[45-47]，B曲线为DMF-Pd NPs样品的FTIR。在3 500 cm^{-1}处的吸收归因于O—H的伸缩振动，应是来源于残留的水。DMF-Pd NPs的FTIR中可以看出，2 930 cm^{-1}处归属于CH_3的对称伸缩振动，2 857 cm^{-1}处归属于C=O的伸缩振动，1 676 cm^{-1}处归属于的C=O的伸缩振动，1 388 cm^{-1}处归属于的HCO—N的弯曲振动，1 257 cm^{-1}处归属于的CN的不对称弯曲振动，观察到在659 cm^{-1}处归属于的HCO—N的伸缩振动，DMF-Pd NPs的FTIR谱带中的大部分谱图都稍微偏移并且比DMF的FTIR的宽度更宽，有可能是由于来自表面Pd原子的电子效应。另外，从谱图还可以看出DMF-Pd NPs样品的红外光谱与DMF红外光谱有较好的重复性，因此我们通过FTIR的测定证实DMF已经附着于Pd NPs的表面，以上结果，我们已经成功制备得到Pd NPs。

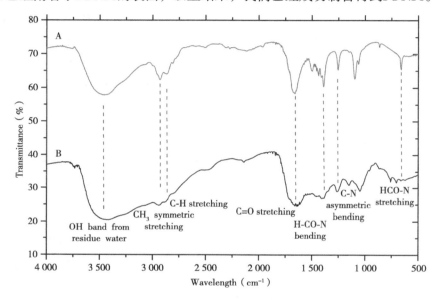

图4.6　DMF（A）和DMF-Pd NPs（B）的红外光谱图

4.3.5　DMF-Pd NPs的TGA

TGA可以用来研究Pd NPs的热稳定性，并对其组分进行预测。在温度升到一定值时，Pd NPs的DMF会脱离钯原子表面而发生热解吸附现象，造成Pd NPs的质量会产生变化。因此可以通过DMF-Pd NPs质量发生的变化的温度与失去的重量，来考察Pd NPs的热稳定信息和其结合的DMF含量。图4.7为DMF-Pd NPs样品的TGA，如图4.7所示，Pd NPs样品的失重比例为57%，其可能来自是Pd NPs样品中DMF的分解和蒸发而造成的，我们设想DMF与Pd NPs结合来自两个层面，包裹在Pd NPs外部的DMF先失重，占样品重量的20%左右，与钯原子结合的DMF失重的重量大约占总重量的37%，因此，推断DMF保护的Pd NPs的平均分子式约为Pd$_{15}$(DMF)$_{13}$。

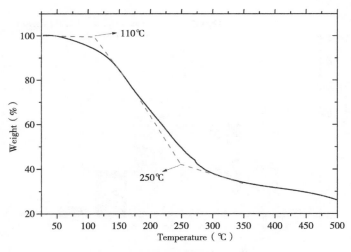

图4.7 DMF-Pd NPs的TGA图

4.3.6 DMF-Pd NPs的TEM和DLS

采用TEM和DLS对DMF-Pd NPs样品的形貌和粒径分布进行表征，图4.8和图4.9分别为反应时间为6 h合成的DMF-Pd NPs的TEM和粒径分布图，从图中可以看出，DMF保护的Pd NPs近似球形，且分散性和粒径分布均一性好，我们选取100个Pd NPs测定其粒径大小，通过计算得出DMF-Pd NPs的平均粒径为（2.20±0.70）mm。图4.10显示了NAC配体与H_2PdCl_4摩尔比为1∶1的NAC-Pd NPs的流体动力学尺寸约为3.4 nm，比TEM的测量结果要大，是因为水溶液中Pd NPs周围有溶剂层的缘故[48]。

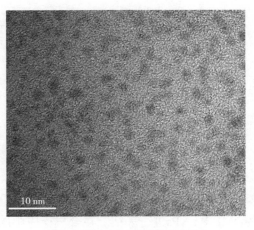

图4.8 DMF-Pd NPs的TEM图

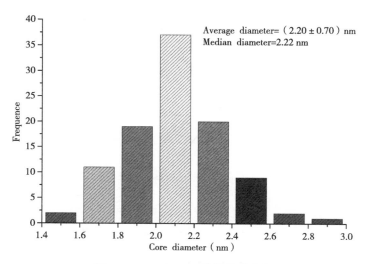

图4.9　DMF-Pd NPs的粒径分布图

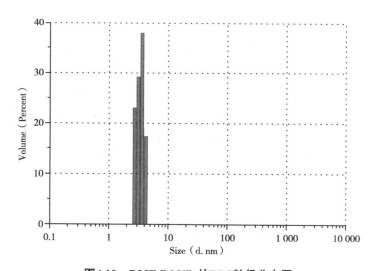

图4.10　DMF-Pd NPs的DLS粒径分布图

4.4　结论

本章采用一步法合成DMF保护的水溶性Pd NPs，通过UV-vis、FL、FTIR、TGA和TEM对合成的DMF-Pd NPs进行表征。在合成反应中DMF同时起着溶剂、配体和还原剂的三重作用，首先，DMF在加热温度高于100℃后会释放的CO可以将钯离子还原为零价钯原子，其次，DMF作为配体和钯离子配位形成配合物，包裹在钯纳米外围，防止Pd NPs的聚集，从而形成DMF-Pd NPs。采用此方法合成的DMF-Pd NPs，具有分散性好，粒径分布均一性好，平均粒径尺寸为（2.20±0.70）nm，且易溶于水和甲醇等有机试剂，有利于在生物和传感器领域的应用。

参考文献

[1] KUMAR S S, KWAK K, LEE D. Electrochemical sensing using quantum-sized gold nanoparticles[J]. Analytical chemistry, 2011, 83(9): 3244-3247.

[2] HASHIMOTO K, ITO K, ISHIMORI Y. Sequence-specific gene detection with a gold electrode modified with DNA probes and an electrochemically active dye[J]. Analytical chemistry, 1994, 66: 3830.

[3] MORENO M, KISSELL L N, JASINSKI J B, et al. Selectivity and reactivity of alkylamine-and alkanethiolate-stabilized Pd and PdAg nanoparticles for hydrogenation and isomerization of allyl alcohol[J]. ACS Catal, 2012, 2(12): 2602-2613.

[4] LI Z P, GAO J, XING X T, et al. Synthesis and characterization of n-alkylamine-stabilized palladium nanoparticles for electrochemical oxidation of methane[J]. The journal of physical chemistry C, 2009, 114(2): 723-733.

[5] LU Y Z, CHEN W. Sub-nanometre sized metal clusters: from synthetic challenges to the unique property discoveries[J]. Chemical society reviews, 2012, 41(9): 3594-3623.

[6] NAKAO Y. Noble metal solid sols in poly(methyl methacrylate)[J]. Journal of colloid and interface science, 1995, 171(2): 386-391.

[7] AYMONIER C, BORTZMEYER D, THOMANN R, et al. Poly(methyl methacrylate)/palladium nanocomposites: synthesis and characterization of the morphological, thermomechanical, and thermal properties[J]. Chemistry of materials, 2003, 15(25): 4874-4878.

[8] WU X L, TAKESHITA S, TADUMI K, et al. Preparation of noble metal/polymer nanocomposites via in situ polymerization and metal complex reduction[J]. Materials chemistry and physics, 2019, 222: 300-308.

[9] DESHMUKH R D, COMPOSTO R J. Surface segregation and formation of silver nanoparticles created in situ in poly(methyl methacrylate) films[J]. Chemistry of materials, 2007, 19(4): 745-754.

[10] 甘颖, 徐国财, 曹震, 等. 纳米Pd/PEG复合材料棒状结构的制备与表征[J]. 化工时刊, 2012, 26(2): 1-4.

[11] 谭德新, 王艳丽, 徐国财. 纳米钯粒子的超声制备与表征[J]. 无机化学学报, 2006, 22(10): 1921-1924.

[12] 邱晓峰, 朱俊杰. 超声化学制备单分散金属纳米钯[J]. 无机化学学报, 2003, 19(7): 766-770.

[13] TU W X, LIU H F. Continuous synthesis of colloidal metal nanoclusters by microwave irradiation[J]. Chemistry of materials, 2000, 12(2): 564-567.

[14] ZHU Y J, CHEN F. Microwave-assisted preparation of inorganic nanostructures in liquid phase[J]. Chemical reviews, 2014, 114(12): 6462-6555.

[15] 刘华蓉, 张志成, 钱逸泰, 等. γ-射线辐照法制备纳米Cu-Pd合金粉末的影响因素[J]. 无机化学学报, 1999, 15(3): 388-392.

[16] GROSS M E, APPELBAUM A, GALLAGHER P K. Laser direct-write metallization in thin palladium acetate films[J]. Journal of applied physics, 1987, 61(4): 1628-1632.

[17] SHAUKAT M S, ZULFIQAR S, SARWAR M I. Incorporation of palladium nanoparticles into aromatic polyamide/clay nanocomposites through facile dry route[J]. Polymer science series B, 2015, 57(4): 380-386.

[18] HAN G Y, GUO B, ZHANG L W, et al. Conductive gold films assembled on electrospun poly(methyl methacrylate) fibrous mats[J]. Advanced materials, 2006, 18(13): 1709-1712.

[19] LIU Z L, FENG Y H, WU X F, et al. Preparation and enhanced electrocatalytic activity of graphene supported palladium nanoparticles with multi-edges and corners[J]. RSC Advances, 2016, 6(101): 98708-98716.

[20] HACHE F, RICARD D, FLYTZANIS C. Optical nonlinearities of small metal

particles: surface-mediated resonance and quantum size effects[J]. Journal of the optical society of America B, 1986, 3(12): 1647-1655.

[21] 杨建新, 王寅, 毛玉荣, 等. Pd/C纳米催化剂的制备及催化Heck反应研究[J]. 化工新型材料, 2014, 42(4): 132-135.

[22] VIET L N, DUC C N, HIROHITO H, et al. Chemical synthesis and characterization of palladium nanoparticles[J]. Advances in natural sciences: nanoscience and nanotechnology, 2010, 1(3): 2043-2047.

[23] CHAN Y N C, CRAIG G S W, SCHROCK R R, et al. Synthesis of palladium and platinum nanoclusters within microphase-separated diblock copolymers[J]. Chemistry of materials, 1992, 4(4): 885-894.

[24] KRÁLIK M, HRONEC M, LORA S, et al. Microporous poly-N,N-dimethylacrylamide-p-styrylsulfonate-methylene bis(acrylamide): a promising support for metal catalysis[J]. Journal of molecular catalysis A: chemical, 1995, 97(3): 145-155.

[25] SADEGHMOGHADDAM E, LAM C, CHOI D, et al. Synthesis and catalytic properties of alkanethiolate-capped Pd nanoparticles generated from sodium S-dodecylthiosulfate[J]. Journal of materials chemistry, 2011, 21(2): 307-312.

[26] CORONADO E, RIBERA A, GARCIA-MARTINEZ J, et al. Synthesis, characterization and magnetism of monodispersed water soluble palladium nanoparticles[J]. Journal of materials chemistry, 2008, 18(46): 5682-5688.

[27] TOSHIMA N, YONEZAWA T. Bimetallic nanoparticles-novel materials for chemical and physical applications[J]. New journal of chemistry, 1998, 22(11): 1179-1201.

[28] XING R, LIU Y M, WU H H, et al. Preparation of active and robust palladium nanoparticle catalysts stabilized by diamine-functionalized mesoporous polymers[J]. Chemical communications, 2008(47): 6297-6299.

[29] 刘兴泉, 唐毅, 戴汉松, 等. N,N-二甲基甲酰胺的生产与应用[J]. 化工科技, 2002, 10(1): 46-49.

[30] LIU X F, LI C H, XU J L, et al. Surfactant-free synthesis and functionalization of highly fluorescent gold quantum dots[J]. The journal of physical chemistry C, 2008, 112(29): 10778-10783.

[31] KAWASAKI H, YAMAMOTO H, FUJIMORI H, et al. Stability of the DMF-protected Au nanoclusters: photochemical, dispersion, and thermal properties[J]. Langmuir, 2009, 26(8): 5926-5933.

[32] KAWASAKI H, YAMAMOTO H, FUJIMORI H, et al. Surfactant-free solution synthesis of fluorescent platinum subnanoclusters[J]. Chemical communications, 2010, 46(21): 3759-3761.

[33] KANG Y J, YE X C, MURRAY C B. Size-and shape-selective synthesis of metal nanocrystals and nanowires using CO as a reducing agent[J]. Angewandte chemie international edition, 2010, 49(35): 6156-6159.

[34] HYOTANISHI M, ISOMURA Y, YAMAMOTO H, et al. Surfactant-free synthesis of palladium nanoclusters for their use in catalytic cross-coupling reactions[J]. Chemical communications, 2011, 47(20): 5750-5752.

[35] XIONG Y, XIA Y. Shape-controlled synthesis of metal nanostructures: the case of palladium[J]. Advanced materials, 2007, 19(20): 3385-3391.

[36] XIONG Y J, CHEN J Y, WILEY B, et al. Size-dependence of surface plasmon resonance and oxidation for Pd nanocubes synthesized via a seed etching process[J]. Nano letters, 2005, 5(7): 1237-1242.

[37] 王旭. 水溶性钯纳米电化学传感器的研究[D]. 太原: 山西大学, 2012.

[38] SHARMA S, KIM B, LEE D. Water-soluble Pd nanoparticles capped with glutathione: synthesis, characterization, and magnetic properties[J]. Langmuir, 2012, 28(45):15958-15965.

[39] EBINA M, IWASA T, HARABUCHI Y, et al. Time-dependent density functional theory study on higher low-lying excited states of $Au_{25}(SR)_{18}$[J]. The journal of physical chemistry C, 2018, 122(7): 4097-4104.

[40] LI M B, TIAN S K, WU Z, et al. Peeling the core-shell Au_{25} nanocluster by reverse ligand-exchange[J]. Chemistry of materials, 2016, 28(4): 1022-1025.

[41] NISHIDA N, SHIBU E S, YAO H, et al. Fluorescent gold nanoparticle superlattices[J]. Advanced materials, 2008, 20(24): 4719-4723.

[42] CHENG PEARL P H, SILVESTER DE, WANG G L, et al. Dynamic and static quenching of fluorescence by 1-4 nm diameter gold monolayer-protected clusters[J]. The journal of physical chemistry B, 2006, 110(10): 4637-4644.

[43] YANG Y Y, CHEN S W. Surface manipulation of the electronic energy of subnanometer-sized gold clusters: an electrochemical and spectroscopic investigation[J]. Nano letters, 2003, 3(1): 75-79.

[44] LINK S, BEEBY A, FITZGERALD S, et al. Visible to infrared luminescence from a 28-atom gold cluster[J]. The journal of physical chemistry B, 2002, 106(13): 3410-3415.

[45] STÅKHANDSKE CHRISTINA M V, MINK J, SANDSTRÖM M, et al. Vibrational spectroscopic and force field studies of N,N-dimethylthioformamide, N,N-dimethylformamide, their deuterated analogues and bis(N,N-dimethylthioformamide)mercury(II) perchlorate[J]. Vibrational spectroscopy, 1997, 14(2): 207-227.

[46] WANG Z X, HUANG B Y, LU Z H, et al. Vibrational spectroscopic studies of interactions between $LiClO_4$ and the plasticizer dimethylformamide[J]. Solid state ionics, 1996, 92(3): 265-271.

[47] QUIST A S, BATES J B, Boyd G E. Raman spectra of molten $NaBF_4$ to 606 °C and 8% NaF-92% to 503 °C[J]. The journal of chemical physics, 1971, 54(11): 4896-4901.

[48] SELIM K M K, KANG I K, GUO H Q. Albumin-conjugated cadmium sulfide nanoparticles and their interaction with KB cells[J]. Macromolecular research, 2009, 17(6): 403-410.

第5章

DMF保护的Pd NPs的色谱分离分析研究

5.1 引言

钯作为贵金属和过渡族金属之一，在现代工业中发挥着重要的作用[1-3]。同时，Pd NPs作为催化剂用于低温还原汽车污染物和有机反应，如Suzuki-Miyaura交叉偶联反应[4]、Heck偶联反应[5]、Stille偶联反应[6]，以及在加氢/脱氢反应[7]和传感器[8]方面也发挥着极其显著的作用。一般而言，通过控制贵金属纳米粒子的粒径可以提高其催化性能。因此，需要了解贵金属纳米粒径如何影响催化性能也就显得特别重要。目前，制备贵金属纳米粒子的方式通常是配体保护还原的金属离子，防止金属原子聚集，而形成稳定的纳米粒子，这种制备贵金属纳米粒子的方法操作简单、重现性好，在反应过程中对周围环境不会产生额外的污染，也是贵金属纳米粒子制备研究未来的发展趋势。Liu等[9]、Kawasaki等[10]和Kawasaki等[11]分别开发了一种新的制备方法，采用DMF作为配体保护的贵金属纳米粒子。在反应过程中，DMF起着溶剂、还原剂和配体的三重作用，且所合成的贵金属纳米粒子具有良好的分散性，可以在有机试剂、水溶液和盐溶液中稳定地存在。但是，研究表明DMF-Au NPs是含有少于20个Au原子的混合物[10]。因此，可以预测DMF保护的贵金属纳米粒子是含有各种贵金属核尺寸或具有相同核心尺寸但具有不同数量的DMF配体。目前，关于DMF-Pd NPs的分离和化学组成的研究未见报道。

实际上，DMF-Pd NPs应该是复杂的混合物，有必要应用有效的分离技术以更好地理解DMF-Pd NPs的化学组成。在众多的分离技术中，高效液相色谱法（HPLC）被认为是一种分离纳米粒子有效的方法。Choi等[12]利用离子对高效液相色谱法对金纳米粒子进行分离，但流动相中的硫酸盐缓冲液配制过程复杂，还会大大缩短色谱柱的使用寿命。Zhang等[13]通过高效液相色谱法成功地实现对金纳米粒子的分离，其采用C18色谱柱，流动相为甲醇和醋酸铵水溶液，研究了甲醇含量对分离的影响以及分离组分的紫外吸收光谱，得到了9种金纳米分离组分，并对分离组分的化学组成进行分析。

电子透射电子显微镜（TEM）被广泛用来金属纳米粒子的表征，但是，使用TEM不能准备确定金属纳米粒子的钯原子和表面保护的配体数量。通过质谱方法可以提供TEM所不能提供的信息[14-17]，此外，质谱技术比只能表征纳米粒子表面配体的NMR时间更快以及需要更少的样品量。

本章采用C18色谱柱（250 mm×4.6 mm，5 μm），流动相为甲醇和水，结合梯度洗脱，利用反向高效液相色谱法分离水溶性DMF-Pd NPs。研究了流动相中不同甲醇含量对DMG-Pd NPs分离的影响，利用UV-vis、FL和MALDI-TOF MS对所分离的Pd NPs组分进行分析。

5.2 实验部分

5.2.1 主要试剂和仪器

$(NH_4)_2PdCl_6$（M=355.22 g/mol，>99.9%）购于Sigma-Aldrich试剂公司；色谱级甲醇和（>99%）均购于Bangkok公司（Thailand）；DHB（98%）购于Sigma公司（St Louis，MO）；氮气（N_2，99.99%，V/V，利泽工业气体经销，太原）；醋酸纤维素过滤膜（0.45 μm）购于天津博纳艾杰尔技术科技有限公司；实验用水为超纯水（Mill-Q Advantage A10超纯水机，默克密理博）；其他所有试剂均为分析纯以上。

Waters Alliance e2695高效液相色谱仪（Millford，MA，美国）包括分离系统，2996二极管阵列检测器，2 475多波长荧光检测器；Bruker Autoflex基质辅助激光解吸电离-飞行时间质谱仪（Bremen，德国）；Zorbax Eclipse Plus-C18色谱柱（250 mm×4.6 mm，5 μm，Agilent Technologies，Stainless steel，CA，美国）；RE 52A真空旋转蒸发仪（上海亚荣生化仪器厂）。

5.2.2 DMF-Pd NPs的制备

将150 μL 0.1 mol/L的$(NH_4)_2PdCl_6$水溶液加入到15 mL 140℃的DMF溶液中，剧烈搅拌6 h后，在真空旋转仪中蒸发多余的溶剂，在室温下，经氮气吹干，得到所需的DMF保护的Pd NPs（DMF-Pd NPs），放干燥器中，备用。

5.2.3 高效液相色谱法的条件

色谱柱：安捷伦Zorbax Eclipse Plus-C18色谱柱（250 mm×4.6 mm，5 μm）；检测器：PDA、荧光检测器；检测波长：200～400 nm（PDA），430 nm（荧光检测器，λ_{ex}=320 nm）；进样量：20 μL；色谱柱温度：（22±1）℃；流速：

0.80 mL/min；流动相：二元流动相，A为色谱级的甲醇，B为二次水。流动相在使用前需用醋酸纤维素（0.45 μm）过滤膜过滤。

梯度洗脱程序如图5.1所示，在0~5 min以3% V/V的流动相A洗脱，5~10 min将流动相A比例线性提高为5% V/V，10~20 min维持流动相A为5% V/V，20~25 min将

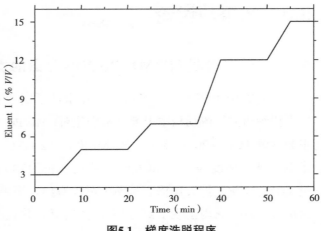

图5.1 梯度洗脱程序

流动相A比例线性提高为7% V/V，25~35 min维持流动相A为7% V/V，35~40 min将流动相A比例线性提高为12% V/V，40~50 min维持流动相A为12% V/V，50~55 min将流动相A比例线性提高为15% V/V，55~60 min维持流动相A为15% V/V。其中Pd NPs经高效液相色谱系统分离，先后由PDA和多波长荧光检测器进行检测，使用紫外-可见信号收集分离的组分。

5.2.4　UV-vis的测量

移取收集的DMF-Pd NPs分离组分溶液至比色皿中，空白为水溶液，用Varian Cary300紫外-可见分光光度计扫描200~500 nm内的吸收光谱。透射电子显微镜（TEM）的表征。

5.2.5　FL的测量

将收集的DMF-Pd NPs分离组分溶液分别置于荧光杯中，在荧光仪上扫描荧光光谱，激发和发射狭缝设为10 nm，实验温度为（22±1）℃。

5.2.6　质谱的测定

将收集的DMF-Pd NPs分离组分甲醇溶液与1.0 mol/L的DHB溶液以1∶1 V/V混合，取混合液4.0 μL滴于质谱样品板上，晾干后，用于MALDI-TOF MS质谱仪进行测定。

5.3 结果与讨论

5.3.1 甲醇含量对DMF-Pd NPs分离的影响

流动相中甲醇的含量可以调节疏水性固定相在色谱柱中的保留时间[18-20]，本章详细研究了流动相中甲醇含量变化对DMF-Pd NPs的分离影响。图5.2为不同甲醇含量（3%~20%，V/V）对DMF-Pd NPs（3.0 mg/mL）样品RP-HPLC分离的影响。洗脱速度为0.80 mL/min，检测波长为300 nm，为了便于比较将甲醇含量（3%，V/V）扩大10倍。从图中可以看出，当甲醇含量为20%（V/V）时，在保留时间10 min内出现了2个很大的洗脱峰，并且在20 min后就再没有色谱峰出现，开始洗脱出来的色谱峰非常的宽大，这应该是许多DMF-Pd NPs色谱峰同时洗脱出来并汇聚起来造成。当甲醇含量为3%（V/V）时，有许多DMF-Pd NPs色谱分离峰被洗脱出来，可以观察到很好的DMF-Pd NPs色谱分离效果，表明DMF保护的Pd NPs是一种混合物样品，其含有不同许多DMF-Pd NPs的洗脱峰。

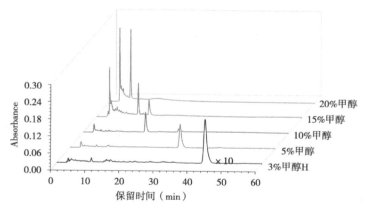

图5.2　甲醇含量对DMF-Pd NPs分离的影响

随着甲醇含量的降低，很好地解决纳米粒子在C18色谱柱中保留时间和洗脱顺序，可以推断DMF-Pd NPs的分离很好地遵循有机物反向高效液相色谱分离的行为[12,21]。图5.3为DMF-Pd NPs与C18色谱柱相互作用示意图。纳米粒子的通过RP-HPLC洗脱顺序与其粒径的大小有关[22-23]，多分散的纳米粒子RP-HPLC分离依赖于非极性固定相和纳米粒子的封闭端配体之间的相互作用[20,24]。当甲醇含量低于≤3%（V/V）时，后面的色谱分离峰有可能被保留在色谱柱内，不能被完全洗脱出来。总之，较早洗脱的色谱峰倾向于用较低的甲醇含量分离，而后面洗脱的色谱峰为避免保留在C18色谱柱上，可以用较高的甲醇含量洗脱，因此，必须应用梯度洗脱程序，在流动相中的甲醇含量从低至高进行分离，以洗脱全部DMF-Pd NPs样品组分。

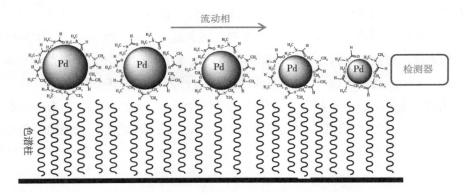

图5.3　DMF-Pd NPs与C18色谱柱相互作用示意图

5.3.2　梯度洗脱对DMF-Pd NPs的分离

RP-HPLC可用于分离纳米材料混合物[13]。图5.4为DMF-Pd NPs样品的色谱分离图，A、B曲线分别为水和DMF-Pd NPs的吸收色谱图，检测波长为300 nm，C曲线为DMF-Pd NPs的FL，激发/发射波长为320/430 nm。

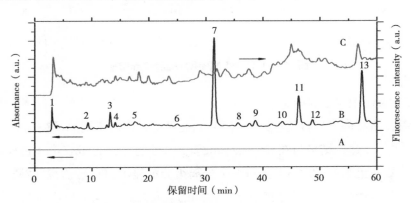

图5.4　DMF-Pd NPs样品的色谱分离图

在图5.4中可以看出，随着甲醇含量的变化，在60 min内，DMF-Pd NPs样品中包含多个紫外吸收和荧光色谱峰，根据紫外色谱峰收集13个具有代表的吸收特征峰，在300 nm紫外吸收下收集分离组分，用于下一步的研究获得DMF-Pd NPs混合物无法得到的性质。参照物水在此梯度洗脱程序中没有任何色谱峰出现。

图5.5为DMF-Pd NPs组分1~6（a）和组分7~13（b）的紫外-吸收光谱图。组分11和组分13为了便于比较扩大了2倍。图中可以看出，分离组分的紫外-吸收光谱具有更多特征的吸收峰。大多组分的在250 nm以下都表现出较强的吸收，并且在波长大于360 nm后没有发现明显的吸收带。其中组分1~4、组分6~7、组分11和组分13表现出更独特的特征特征吸收峰。组分7、组分11和组分13分别在310 nm、290 nm、300 nm处有吸收峰，这些分离组分特征峰隐藏于DMF-Pd NPs混合物紫外吸收图谱。实际上，分离组分的紫外光谱吸收峰影响合成产物Pd NPs紫外光谱中250~400 nm处的宽吸收带，详见第4章图4.3 DMF-Pd NPs的紫外光谱图。所有分离组分的吸收峰没有特有的SPR，表明分离的DMF-Pd NPs的粒径都小于3 nm[25-27]。

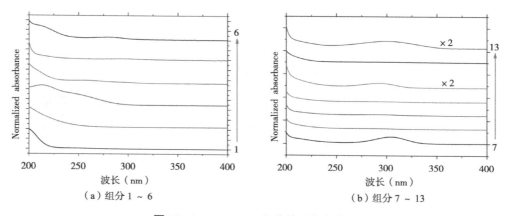

图5.5　DMF-Pd NPs组分的吸收光谱图

图5.6为DMF-Pd NPs组分1~6（a）和7~13（b）的荧光光谱图。当λ_{ex}为300 nm时，所有的发射峰均在350~500 nm处显示，归因于钯原子在4d带和5sp带的跃迁[23,28-29]。这一结果与在320/422 nm的$\lambda_{ex}/\lambda_{em}$处的合成的DMF-PdNPs样品的荧光光谱一致（见第4章图4.5）。每一个分离组分都显示出自己的特征发射带，这是从制备的DMF-PdNPs混合物中不可得到的。总之，不同分离组分具有不同的光学性质归因于Pd NPs表面结合DMF的数量有关。色谱分离与吸收和荧光光谱结合提供了评估分离的Pd NPs样品的光谱特征的重要且有力的工具。此外，收集的Pd NPs组分也可以通过其他技术进行表征和研究。

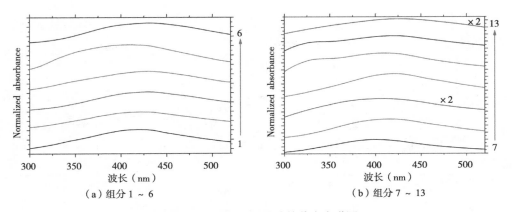

图5.6 DMF-Pd NPs组分的荧光光谱图

5.3.3 质谱对DMF-Pd NPs分离组分的分析

一直以来，如何确定金属纳米粒子的确切质量（包括钯原子和表面保护配体的数量）是一项挑战，自从Whetten等[14]使用MALDI-TOF MS分析金属纳米粒子后，发现MS是一种可靠的技术来测量小尺寸纳米粒子的分子量。从那以后，MS方法几乎在所有金属纳米粒子的结构分析中被广泛采用。

图5.7为使用MALDI-TOF MS对DMF-Pd NPs组分1~13分析的质谱图。Pd NPs经MS激光辐照后碎裂，从而明确Pd NPs组分的分子量峰。我们已经最小化了MS的激光辐照功率，然而Pd NPs组分的碎裂是不可避免的。但是，当激光辐照功率太低时，我们并没有观察到分子量峰。所有m/z峰都低于3 000 Da，并且所有峰分配都在MALDI-TOF MS的m/z的精度（±1~2 m/z）内，最大分子量基线的信噪比≥3。每个部分显示一系列主要分子量间距，表明碎片化的Pd NPs离子。图中用$Pd_x(DMF)_y$的形式来表示，其中x代表分离组分中钯原子数，y代表分离组分中DMF的配体数。表5.1为质谱峰$Pd_x(DMF)_y$的钯原子和NAC的数目分配。组分1~13最大分子量分别代表$Pd_{14}(DMF)_8$、$Pd_{10}(DMF)_8$、$Pd_{12}(DMF)_{11}$、$Pd_{14}(DMF)_8$、$Pd_{14}(DMF)_{10}$、$Pd_{14}(DMF)_{12}$、$Pd_{15}(DMF)_{12}$、$Pd_{16}(DMF)_{15}$、$Pd_{16}(DMF)_{15}$、$Pd_{16}(DMF)_{15}$、$Pd_{20}(DMF)_9$、$Pd_{17}(DMF)_{13}$和$Pd_{20}(DMF)_9$。组分8-10具有相同的分子式$Pd_{16}(DMF)_{15}$，组分11、组分13既具有相同的分子式$Pd_{20}(DMF)_9$，但从C18色谱柱洗脱的先后顺序不同，归因于DMF配体与Pd NPs的配位差异[30]。

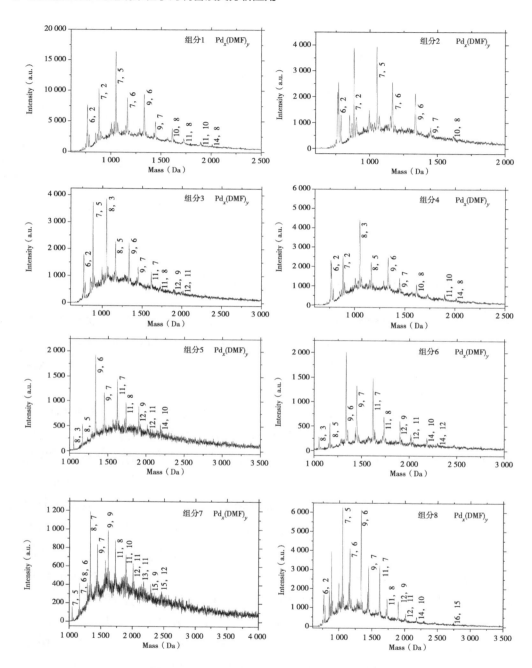

图5.7 DMF-Pd NPs分离组分1~13的质谱图

第 5 章 DMF 保护的 Pd NPs 的色谱分离分析研究

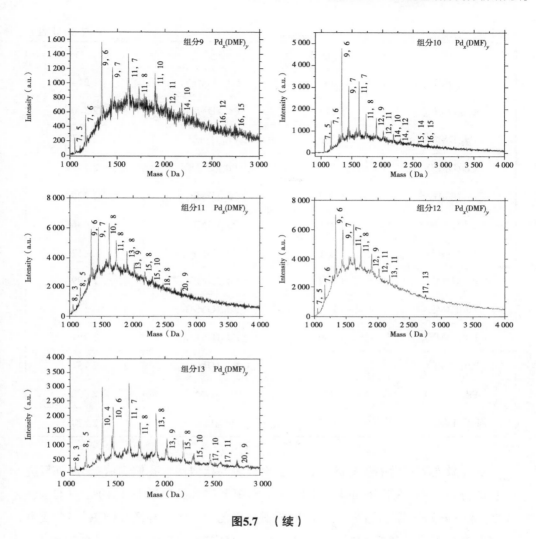

图5.7 （续）

表5.1 质谱峰 $Pd_x(DMF)_y$ 的钯原子和 NAC 的数目分配

分子式	m/z（Da）	分子式	m/z（Da）
$Pd_6(DMF)_2$	782	$Pd_{13}(DMF)_8$	1 962
$Pd_7(DMF)_2$	888	$Pd_{13}(DMF)_9$	2 035
$Pd_7(DMF)_5$	1 107	$Pd_{13}(DMF)_{11}$	2 181
$Pd_7(DMF)_6$	1 180	$Pd_{14}(DMF)_8$	2 068

（续表）

分子式	m/z（Da）	分子式	m/z（Da）
$Pd_8(DMF)_3$	1 067	$Pd_{14}(DMF)_{10}$	2 214
$Pd_8(DMF)_5$	1 213	$Pd_{14}(DMF)_{12}$	2 360
$Pd_8(DMF)_6$	1 286	$Pd_{15}(DMF)_8$	2 174
$Pd_8(DMF)_7$	1 359	$Pd_{15}(DMF)_9$	2 247
$Pd_9(DMF)_6$	1 392	$Pd_{15}(DMF)_{10}$	2 320
$Pd_9(DMF)_7$	1 465	$Pd_{15}(DMF)_{12}$	2 466
$Pd_9(DMF)_9$	1 611	$Pd_{15}(DMF)_{14}$	2 612
$Pd_{10}(DMF)_8$	1 644	$Pd_{16}(DMF)_{11}$	2 499
$Pd_{11}(DMF)_7$	1 677	$Pd_{16}(DMF)_{12}$	2 572
$Pd_{11}(DMF)_8$	1 750	$Pd_{16}(DMF)_{15}$	2 791
$Pd_{11}(DMF)_{10}$	1 896	$Pd_{17}(DMF)_{13}$	2 751
$Pd_{12}(DMF)_9$	1 929	$Pd_{18}(DMF)_8$	2 492
$Pd_{12}(DMF)_{11}$	2 075	$Pd_{20}(DMF)_9$	2 777

在经激光辐照后的组分碎片，显示出很多的相似性，我们选取组分1进行进一步的分析。图5.8组分1的质谱图（A）和矩形放大的质谱图（B）、（C）、（D）和（E）分别为质谱（B）不同质谱范围的放大图。在图5.8（B）3个质谱峰之间有分子量差值为285（±1~2 m/z），对应着1个DMF配体（M=73 m/z）和2个Pd原子（106×2=206 m/z）。从图5.8（C）、（D）和（E）可以看出，在每组碎片中也发现小的m/z间距，这系列每一个小峰碎片离子都有的来自同一组的分子量损失。这些不同小质谱峰为—CH_3、—OH或H_2O分子、2个—CH_3或质子化的HC=O$^+$H、N（CH_3）$_2$和2个—CHO分别对应15、17或18、30、44和58（m/z），而相同的小质谱峰同样在DMF质谱中会被找到。通过质谱分析，确定了DMF-Pd NPs样品经RP-HPLC分离组分的化学组成，证实了DMF-Pd NPs具有不同分子量的各种类型钯纳米的复杂混合物。

第 5 章 DMF 保护的 Pd NPs 的色谱分离分析研究

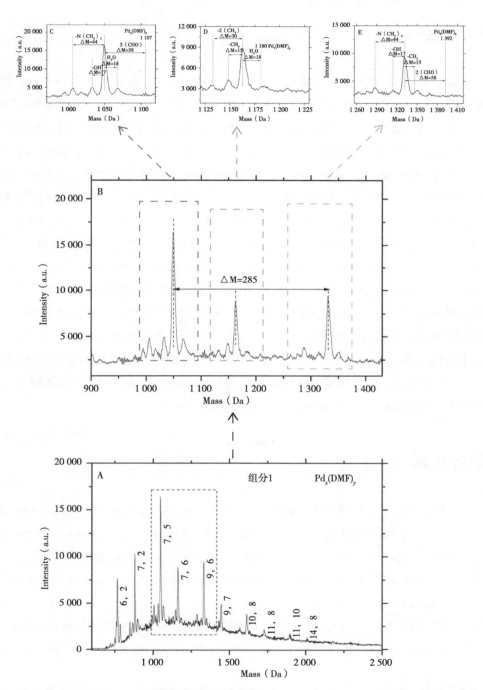

图5.8 组分1的质谱图（A）和矩形放大的质谱图（B）、（C）、（D）和（E）分别为质谱（B）不同质谱范围的放大图

5.4 结论

本章采用C18色谱柱（250 mm×4.6 mm，5 μm），流动相为甲醇和水，结合梯度洗脱，利用RP-HPLC分离水溶性DMF-Pd NPs。研究了流动相中不同甲醇含量对DMF-Pd NPs分离的影响，利用UV-vis、FL和MALDI-TOF MS。研究表明，通过RP-HPLC至少分离出的13种DMF-Pd NPs组分，每一个分离组分都有自己特征的UV-vis和FL，这些性质是在研究DMF-Pd NPs混合物时无法获知的。采用MALDI-TOF-MS对DMF-Pd NPs分离组分进行分析，可显示每一组分的粒子特性，得到分离组分的化学组成，分别为$Pd_{14}(DMF)_8$、$Pd_{10}(DMF)_8$、$Pd_{12}(DMF)_{11}$、$Pd_{14}(DMF)_8$、$Pd_{14}(DMF)_{10}$、$Pd_{14}(DMF)_{12}$、$Pd_{15}(DMF)_{12}$、$Pd_{16}(DMF)_{15}$、$Pd_{16}(DMF)_{15}$、$Pd_{16}(DMF)_{15}$、$Pd_{20}(DMF)_9$、$Pd_{17}(DMF)_{13}$和$Pd_{20}(DMF)_9$。其中组分8-10具有相同的分子式$Pd_{16}(DMF)_{15}$，组分11、组分13既具有相同的分子式$Pd_{20}(DMF)_9$，但是从C18色谱柱洗脱的先后顺序不同，归因于DMF配体与Pd NPs的配位差异。以前没有报道过利用RP-HPLC使用甲醇和纯水作为流动相分离小尺寸的Pd NPs，在此，我们已成功将RP-HPLC方法应用于在所制备的DMF-Pd NPs产物的分离，本章的研究为纳米材料范围内的Pd NPs及其他贵金属纳米粒子的分离、吸收和FL性质研究提供了一定的基础。

参考文献

[1] HUANG M, CHENG J H, TAO X C, et al. Application of suzuki cross-coupling reaction catalyzed by ligandless palladium chloride in the synthesis of liquid crystals[J]. Synthetic communications, 2007, 37(13): 2203-2208.

[2] ANTLER M. The application of palladium in electronic connectors[J]. Platinum metals review, 1982, 26(3): 106-117.

[3] TUNGLER A, TARNAI T, HEGEDÛS L, et al. Palladium-mediated heterogeneous catalytic hydrogenations[J]. Platinum metals review, 1998, 42(3): 108-115.

[4] MAGDESIEVA T V, NIKITIN O M, LEVITSKY O A, et al. Polypyrrole-palladium nanoparticles composite as efficient catalyst for Suzuki-Miyaura

coupling[J]. Journal of molecular catalysis A: chemical, 2012, 353-354: 50-57.

[5] KIM H S, KIM J D, CHOI H C, et al. UV-irradiation-mediated palladium nanoparticle catalytic system: heck and decarboxylative coupling reactions[J]. Molecular catalysis, 2017, 441: 21-27.

[6] HERAVI M M, MOHAMMADKHANI L. Recent applications of stille reaction in total synthesis of natural products: an update[J]. Journal of organometallic chemistry, 2018, 869: 106-200.

[7] TUNC N, ABAY B, RAKAP M. Hydrogen generation from hydrolytic dehydrogenation of hydrazine borane by poly(N-vinyl-2-pyrrolidone)-stabilized palladium nanoparticles[J]. Journal of power sources, 2015, 299: 403-407.

[8] ALFANO B, MASSERA E, POLICHETTI T, et al. Effect of palladium nanoparticle functionalization on the hydrogen gas sensing of graphene based chemi-resistive devices[J]. Sensors and actuators B: chemical, 2017, 253: 1163-1169.

[9] LIU X F, LI C H, XU J L, et al. Surfactant-free synthesis and functionalization of highly fluorescent gold quantum dots[J]. Journal physical chemistry C, 2008, 112(29): 10778-10783.

[10] KAWASAKI H, YAMAMOTO H, FUJIMORI H, et al. Stability of the DMF-protected Au nanoclusters: photochemical, dispersion, and thermal properties[J]. Langmuir, 2009, 26(8): 5926-5933.

[11] KAWASAKI H, YAMAMOTO H, FUJIMORI H, et al. Surfactant-free solution synthesis of fluorescent platinum subnanoclusters[J]. Chemical communications, 2010, 46(21): 3759-3761.

[12] CHOI MARTIN M F, DOUGLAS A D, MURRAY R W. Ion-pair chromatographic separation of water-soluble gold monolayer-protected clusters[J]. Analytical chemistry, 2006, 78(8): 2779-2785.

[13] ZHANG Y, HU Q, PAAU M C, et al. Probing histidine-stabilized gold nanoclusters product by high-performance liquid chromatography and mass spectrometry[J]. The journal of physical chemistry C, 2013, 117(36): 18697-18708.

[14] WHETTEN R L, KHOURY J T, ALVAREZ M M, et al. Nanocrystal gold molecules[J]. Advanced materials, 1996, 8(5): 428-433.

[15] KOUCHI H, KAWASAKI H, ARAKAWA R. A new matrix of MALDI-TOF MS for the analysis of thiolate-protected gold clusters[J]. Analytical methods, 2012, 4(11): 3600-3603.

[16] YANG X P, SU Y, PAAU M C, et al. Mass spectrometric identification of water-soluble gold nanocluster fractions from sequential size-selective precipitation[J]. Analytical chemistry, 2012, 84(3): 1765-1771.

[17] ARNOLD RANDY J, REILLY JAMES P. High-resolution time-of-flight mass spectra of alkanethiolate-coated gold nanocrystals[J]. Journal of the American chemical society, 1998, 120(7): 1528-1532.

[18] HU Q, GONG X J, LIU L Z, et al. Characterization and analytical separation of fluorescent carbon nanodots[J]. Journal of nanomaterials, 2017, 2017: 23.

[19] HU Q, PAAU M C, CHOI MARTIN M F, et al. Better understanding of carbon nanoparticles via high-performance liquid chromatography-fluorescence detection and mass spectrometry[J]. Electrophoresis, 2014, 35(17): 2454-2462.

[20] ZHANG L, HU Q, LI Z P, et al. Chromatographic separation and mass spectrometric analysis of N-acetyl-L-cysteine-protected palladium nanoparticles[J]. Analytical methods, 2017, 9(31): 4539-4546.

[21] GONG X J, HU Q, CHIN P M, et al. High-performance liquid chromatographic and mass spectrometric analysis of fluorescent carbon nanodots[J]. Talanta, 2014, 129: 529-538.

[22] WOLFE R L, MURRAY R W. Analytical evidence for the monolayer-protected cluster $Au_{225}[(S(CH_2)_5CH_3)]_{75}$[J]. Analytical chemistry, 2006, 78(4): 1167-1173.

[23] ZHANG Y, SHUANG S M, DONG C, et al. Application of HPLC and MALDI-TOF MS for studying as-synthesized ligand-protected gold nanoclusters products[J]. Analytical chemistry, 2009, 81(4): 1676-1685.

[24] KRULL I S, SWARTZ M E. Books: for every chromatographer[J]. Analytical chemistry, 1997, 69(23): 740A.

[25] XIONG Y, XIA Y. Shape-controlled synthesis of metal nanostructures: the case

of palladium[J]. Advanced materials, 2007, 19(20): 3385-3391.

[26] JIMENEZ V L, GEORGANOPOULOU D G, WHITE R J, et al. Hexanethiolate monolayer protected 38 gold atom cluster[J]. Langmuir, 2004, 20(16): 6864-6870.

[27] DANIEL M C, ASTRUC D. Gold nanoparticles: assembly, supramolecular chemistry, quantum-size-related properties, and applications toward biology, catalysis, and nanotechnology[J]. Chemical reviews, 2004, 104(1): 293-346.

[28] LINK S, BEEBY A, FITZGERALD S, et al. Visible to infrared luminescence from a 28-atom gold cluster[J]. The journal of physical chemistry B, 2002, 106(13): 3410-3415.

[29] WANG G L, HUANG T, MURRAY R W, et al. Near-IR luminescence of monolayer-protected metal clusters[J]. Journal of the American chemical society, 2005, 127(3): 812-813.

[30] NIIHORI Y, MATSUZAKI M, PRADEEP T, et al. Separation of precise compositions of noble metal clusters protected with mixed ligands[J]. Journal of the American chemical society, 2013, 135(13): 4946-4949.

第6章

DMF-Pd NPs修饰的玻碳电极对Cu^{2+}的电化学检测

6.1 引言

铜作为人体必不可少的微量元素之一，能够以多种形式存在于人类的生存环境中，但随着工农业的发展，铜离子通过各种途径进入人体内，过量铜的摄入会沉积在人脑、胰腺、肝脏等，导致人体内的蛋白质结构发生不可逆的改变，影响生理功能，从而会导致严重的疾病和后果[1-4]。此外，铜离子污染造成对生态环境的破坏也十分严重，铜及其化合物在我国是优先控制的污染物而被列入环境污染物的"黑名单"，因此，对铜离子快速、准确、高效的测定分析方法的建立也是势在必行。目前为止，常用的铜离子测定方法主要有UV-vis[5]、原子光谱法[6-7]、FL[8-10]、化学发光法[11]和电感耦合等离子体质谱法[12-14]等，但是，上述的方法在测定铜离子时，样品前处理过程繁琐、操作步骤复杂、仪器设备成本高，相比而言，电化学方法因具有操作简单、选择性好、测定速度快、成本低廉、便于携带和微型化等优点，引起人们越来越多的关注[15]。

进入21世纪，随着纳米科技的快速发展，贵金属纳米粒子表现出特有的光学、电学、催化等性能[16-18]，特别是电化学测定方面尤其备受科研工作者的关注[19-20]。例如，杨帅等[21]采用Pd NPs/MWNTs修饰玻碳电极，实现了对金属离子Cr^{6+}快速、简便的电化学检测，该方法具有检出限低和选择性好等优点。朱伟明等[22]合成无负载的金-钯合金纳米颗粒，该纳米颗粒合成方法简便，性质稳定，该纳米粒子修饰的玻碳电极催化性能好、成本低、重现性好，检出限达到8×10^{-7} mol/L，可用于快速检测食品中的H_2O_2残留量。Kumar等[23]制备了巯基配体保护的金纳米粒子，并将其修饰于电极，测定样品中抗坏血酸和尿酸，研究表明，金纳米粒子在修饰电极中起着氧化还原中介与电子导体的双重作用。Abollino等[24]将制备的Au NPs修饰于玻碳电极上，通过金-汞原子之间的化学亲和作用，实现了对金属Hg^{2+}的测定。Ahmad等[25]将制备的Au NPs修饰于玻碳电极，通过循环伏安法测试后，研究发现该修饰电极对邻苯二酚有着卓越的催化氧化性能。Sang等[26]将制备的银纳米与还原氧化石墨烯复合纳米材料，修饰于玻碳电极，通过与还原氧化石墨烯修饰电极比较，发现Ag NPs/RGO修饰电极的选择性和催化活性更好，并可同时测定Pb^{2+}、Hg^{2+}、Cr^{2+}等重金属离子。Wu等[27]将制备的石墨烯/Pd NPs修饰电极，用于检测消毒防腐药剂三氯生，该电极具有灵敏度高、操作简单、快速便捷和检出限低的特点。Zhang等[28]将制备的金-Pd NPs

用于对金属离子As^{3+}的电化学响应，利用金属纳米粒子大的比表面积，从而显著提高金属离子As^{3+}在测定中的响应信号，该方法具有较高的灵敏度，检测限为2.4×10^{-2} μg/L。因此，用电化学方法制备的贵金属纳米粒子修饰电极测定环境中有害金属离子有着先天特有的优势和广阔的实际应用前景[29]。

本章利用DMF-Pd NPs良好的稳定性和溶解性，采用滴涂法制备DMF-Pd NPs修饰玻碳电极，并在修饰电极电化学表征的基础上，通过优化DMF-Pd NPs/GC修饰电极对Cu^{2+}的电化学响应性能，建立了一种基于DMF-Pd NPs/GC修饰电极对Cu^{2+}的测定方法，该修饰电极具有灵敏度高、选择性好等优点，因此，有望实现对环境水样中Cu^{2+}实时快速的测定。

6.2　实验部分

6.2.1　主要试剂和仪器

仪器：CHI660C电化学仪（上海市辰华仪器有限公司出产）；Mill-Q Advantage A10超纯水机（默克密理博）；ME 204万分之一天平（METLER TLIEDO，上海）；GVS-2L超声波清洗仪（深圳市够威科技有限公司）；JB-A型智能控温磁力搅拌器（上海雷磁创意仪器仪表有限责任公司）；RE 52A真空旋转蒸发仪（上海亚荣生化仪器厂）。

试剂：Nafion（25%，V/V）为上海阿拉丁生化科技股份有限公司；$(NH_4)_2PdCl_6$（M=355.22 g/mol，>99.9%）购于Sigma-Aldrich试剂公司；DMF（>99%）购于Bangkok公司（Thailand）；硫酸铜（$CuSO_4 \cdot 5H_2O$，M=249.68 g/mol，分析纯）为北京化工厂；硫酸锌（$ZnSO_4 \cdot 7H_2O$，M=287.54 g/mol，分析纯）为天津科密欧化学试剂有限公司；硫酸锰（$MnSO_4 \cdot H_2O$，M=169.01 g/mol，分析纯）为天津化学试剂有限公司；氯化钴（$CoCl_2 \cdot 6H_2O$，M=237.93 g/mol，分析纯）为上海试剂厂；硫酸镍（$MgSO_4 \cdot 7H_2O$，M=246.37 g/mol，分析纯）为天津试剂一厂；硫酸亚铁（$FeSO_4 \cdot 7H_2O$，M=278.02 g/mol，分析纯）为北京化工厂；氯化汞（$HgCl_2 \cdot H_2O$，M=271.50 g/mol，分析纯）为上海试剂二厂；氯化铅（$PbCl_2 \cdot H_2O$，M=278.12 g/mol，分析纯）为天津试剂厂；硫酸镉（$3CdSO_4 \cdot 8H_2O$，M=769.52 g/mol，分析纯）为北京化工厂；其余所用试剂均

为分析纯。实验用水为超纯水。

6.2.2 DMF-Pd NPs的制备

将150 μL 0.1 mol/L的（NH_4）$_2PdCl_6$水溶液加入到15 mL 140℃的DMF溶液中，剧烈搅拌6 h后，在真空旋转仪中蒸发多余的溶剂，在室温下，经氮气吹干，得到所需的DMF保护的DMF-Pd NPs，放干燥器中，备用。

6.2.3 修饰电极的制备

首先，将GC在金相砂纸上研磨；其次，用0.5 μm的Al_2O_3粉末对电极表面抛光打磨后；再次，依次在超纯水、丙酮、超纯水中超声2 min、30 s、2 min，在GC电极表面滴加1 mg/mL DMF-Pd NPs溶液2~20 μL，在室温条件下，经自然晾干；最后，再滴加2 μL 2%的Nafion溶液，经红外灯烘干1 min后，即制备出DMF-Pd NPs修饰的电极（DMF-Pd NPs/GC）。

6.2.4 电化学试验方法

采用三电极体系测量：其中饱和甘汞电极（SCE）为参比电极，铂电极为对电极，DMF-Pd NPs/GC电极和GC电极为工作电极。采用循环伏安法检测Cu^{2+}的电化学行为。

6.3 结果与讨论

6.3.1 DMF-Pd NPs的合成

图6.1为DMF-Pd NPs的合成示意图。参照本书中4.3.1的合成讨论以及根据我们之前的试验研究经验[30]，在反应中DMF同时起着反应溶液、配体和还原剂的作用。首先，DMF作为配体和钯离子配位形成配合物；其次，当DMF的加热温度高于100℃后会释放出二甲胺和CO，其中CO是钯离子形成Pd NPs的还原剂，从而形成DMF-Pd NPs[31]。

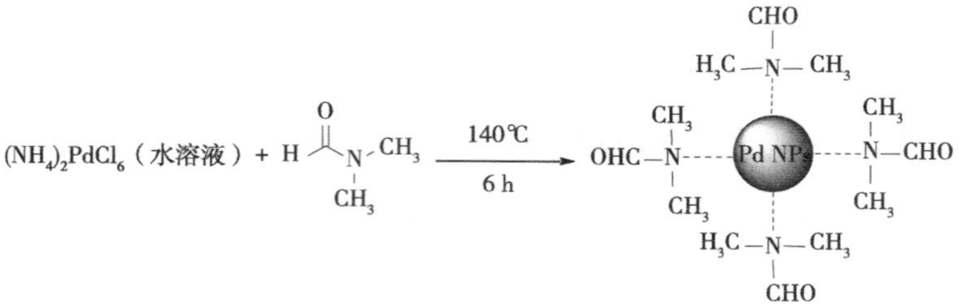

$(NH_4)_2PdCl_6$（水溶液）+ HCON(CH_3)_2 $\xrightarrow[6\text{ h}]{140\text{℃}}$ DMF-Pd NPs

图6.1　DMF-Pd NPs合成示意图

6.3.2　修饰电极在电解质溶液中的循环伏安行为

如图6.2所示为DMF-Pd NPs/GC电极和GC电极在0.1 mol/L的硫酸溶液的循环伏安图，其扫描电位范围为−0.0～1.8 V，扫描速度为50 mV/s。与GC电极A相比较，DMF-Pd NPs/GC电极B的钯纳米的循环伏安氧化峰在1.3 V左右，还原峰在0.7 V左右，且其峰电流值最大，催化能力显著提高，表明将DMF-Pd NPs修饰于玻碳电极时，可以显著提高电极的催化活性。

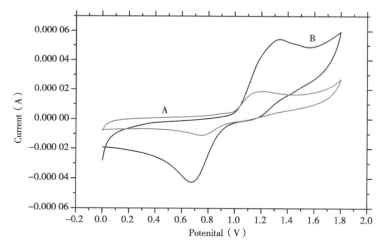

图6.2　GC电极A和DMF-Pd NPs/GC电极B在电解液中的循环伏安图

6.3.3　电极修饰量的优化

修饰电极的电化学氧化还原活性，以及对被检测物质的灵敏度与电极修饰材料的使用量有一定的关系，因此，我们考察DMF-Pd NPs在电极上的修饰量。结果如图6.3所示，修饰量在2~20 μg内，当修饰量达到10 μg的时，修饰电极对浓度为5×10^{-5} mol/L的Cu^{2+}电化学响应信号最好，因此，我们的实验中选择DMF-Pd NPs在电极上的修饰量为10 μg进行检测。

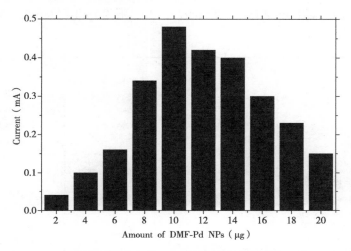

图6.3 DMF-Pd NPs修饰量对电极的影响

6.3.4 电极扫描速度和扫描圈数的优化

研究扫描圈数和扫描速度对修饰电极检测铜离子的影响。实验中以扫描速率5 mV/s、50 mV/s、75 mV/s、100 mV/s对修饰电极响应灵敏度进行了考察，结果表明，扫描速度对修饰电极灵敏度有较小的影响。因此，实验中基于省时和循环伏安曲线完美的基础上选择扫描速度为50 mV/s。扫描圈数实验中以不同的扫描圈数（1圈、2圈、3圈、5圈）进行稳定性条件优化，扫描速率为50 mV/s，扫描圈数为2圈时，修饰电极的检测信号完全达到了稳定。这是因为被检测离子在溶液中向电极表面扩散和吸附过程存在一个不稳地结合过程。因此，在本实验中选择扫速50 mV/s，扫描圈数为2圈进行检测。

6.3.5 DMF-Pd NPs修饰电极对Cu^{2+}的检测

如图6.4所示，采用循环伏安法对不同浓度的Cu^{2+}进行电化学检测，其中电位范围为-0.4~1.4 V，扫描速度为50 mV/s，通过依次向5 mL 0.1 mol/L H_2SO_4电解质溶液中加入不同浓度的$CuSO_4$溶液，发现DMF-Pd NPs修饰的玻碳电极，对不同浓度的Cu^{2+}都有较好的识别能力。随着铜离子浓度的不断增加，氧化峰电流也逐渐增大，从图6.5中可以得出，在$（4×10^{-7}）$~$（5×10^{-5}）$ mol/L浓度范围内铜离子表现出较好的线性关系。其线性方程为：Ip（μA）=5.572 7×10^{-7}+2.244 4×10^{-4}C（μmol/L），R^2=0.970，检出限为$5×10^{-7}$ mol/L（S/N=3）。

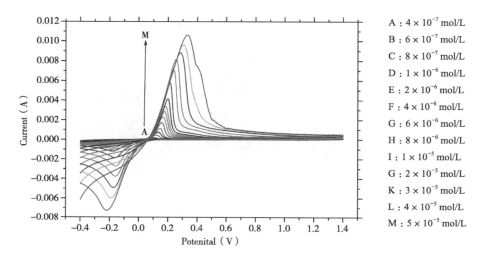

图6.4　不同浓度的Cu^{2+}在电解液中的循环伏安图

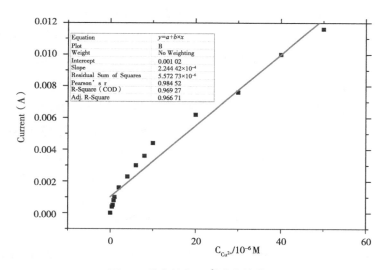

图6.5　峰电流与Cu^{2+}浓度的关系

6.3.6 DMF-Pd NPs修饰电极对金属铜离子的选择性

如图6.6所示,在0.1 mol/L H_2SO_4电解液中加入浓度为$5×10^{-5}$ mol/L的Cu^{2+}进行循环伏安测定,再分别加入同等浓度的金属离子进行对比测定,以评价其方法的选择性。在电解液中,DMF-Pd NPs/GC修饰电极对Cd^{2+}、Zn^{2+}、Mn^{2+}、Co^{2+}、Mg^{2+}、Fe^{2+}、Hg^{2+}、Pb^{2+}等常见的重金属污染物离子几乎没有电化学响应行为,而只是对Cu^{2+}表现出较好的响应性能,说明DMF-PdNPs/GC修饰电极对金属Cu^{2+}良好的选择性。

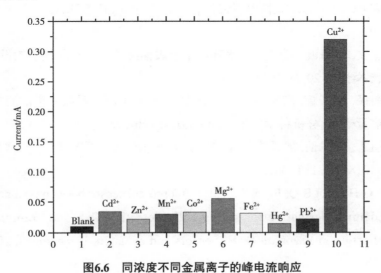

图6.6 同浓度不同金属离子的峰电流响应

6.4 结论

本章采用滴涂法制备DMF-Pd NPs修饰玻碳电极,并在修饰电极电化学表征的基础上,通过优化DMF-Pd NPs/GC修饰电极对Cu^{2+}的电化学响应性能,建立了一种基于DMF-Pd NPs/GC修饰电极对Cu^{2+}的测定方法,线性范围为($4×10^{-7}$)~($5×10^{-5}$)mol/L浓度,其线性方程为:Ip(μA)=$5.5727×10^{-7}+2.2444×10^{-4}$C(μmol/L),$R^2$=0.970,检出限为$5×10^{-7}$ mol/L(S/N=3),本章所研制的DMF-Pd NPs/GC修饰电极具有成本低、操作简单、灵敏度高和选择性好的优点,因此,该电极对于测定环境水样中的Cu^{2+}具有潜在应用价值。

参考文献

[1] 李征. 重金属污染水体的环境保护处理技术研究[J]. 环境与发展, 2018, 30(11): 110-111.

[2] 钟燕, 周洁丹, 刘英菊, 等. 新型巯基化合物自组装膜修饰电极对铜离子的检测[J]. 分析测试学报, 2009, 28(9): 1031-1034, 1039.

[3] 窦薛楷. 浅谈铜的污染及危害[J]. 科技经济导刊, 2017(8): 126-127.

[4] 朱朝晖. 芦丁修饰碳糊电极测定微量的铜离子[J]. 安徽化工, 2006, 32(1): 68-70.

[5] 李志文, 李海波, 万浩, 等. 水环境重金属铜离子光学检测仪器的设计[J]. 传感技术学报, 2016(1): 146-149.

[6] 卢爱民, 柴辛娜, 高宏宇, 等. 溶剂萃取-石墨炉原子吸收光谱法测定水样中的痕量铅[J]. 分析科学学报, 2006, 22(2): 190-192.

[7] 周道志, 曾凤仙. 原子吸收光谱法测定蔬菜中铁、铜的含量[J]. 食品安全导刊, 2018(27): 110-111.

[8] DING H C, LI B Q, PU S Z, et al. A fluorescent sensor based on a diarylethene-rhodamine derivative for sequentially detecting Cu^{2+} and arginine and its application in keypad lock[J]. Sensors and actuators B: chemical, 2017, 247: 26-35.

[9] MA L L, LIU G, PU S Z, et al. A highly selective fluorescent chemosensor for Cu^{2+} based on a new diarylethene with triazole-linked fluorescein[J]. Tetrahedron, 2016, 72(7): 985-991.

[10] DONG J Y, HU J F, BAIGUDE H, et al. A novel ferrocenyl-naphthalimide as a multichannel probe for the detection of Cu(ii) and Hg(ii) in aqueous media and living cells[J]. Dalton transactions, 2018, 47(2): 314-322.

[11] 屈颖娟, 翟云会, 王亚妮. 水中铜离子的高选择性检测方法[J]. 光谱实验室, 2013(5): 249-252.

[12] CHAI X, ZHOU X, ZHU A, et al. A two-channel ratiometric electrochemical biosensor for in vivo monitoring of copper ions in a rat brain using gold truncated octahedral microcages[J]. Angewandte chemie, 2013, 52(31): 8129-8133.

[13] 周瑞妮, 庞恩, 李庚. 质控图在电感耦合等离子体质谱法测定食品中9种重金属中的应用[J]. 食品安全质量检测学报, 2018, 9(19): 5242-5246.

[14] BORA T, AKSOY A, TUNAY Z, et al. Determination of trace elements in illicit spice samples by using ICP-MS[J]. Microchemical journal, 2015, 123: 179-184.

[15] LI M, GOU H L, AL-OGAIDI I, et al. Nanostructured sensors for detection of heavy metals: a review[J]. ACS Sustainable chemistry & engineering, 2013, 1(7): 713-723.

[16] LI Z P, GAO J, XING X T, et al. Synthesis and characterization of n-alkylamine-stabilized palladium nanoparticles for electrochemical oxidation of methane[J]. The journal of physical chemistry C, 2009, 114(2): 723-733.

[17] AIKEN III J D, FINKE R G. A review of modern transition-metal nanoclusters: their synthesis, characterization, and applications in catalysis[J]. Journal of molecular catalysis A: chemical, 1999, 145(1-2): 1-44.

[18] HUANG J L, LIN L Q, SUN D H, et al. Bio-inspired synthesis of metal nanomaterials and applications[J]. Chemical society reviews, 2015, 44(17): 6330-6374.

[19] ALLERSTON L K, REES N V. Nanoparticle impacts in innovative electrochemistry[J]. Current opinion in electrochemistry, 2018, 10: 31-36.

[20] 荣晓娇, 石磊, 丁士明, 等. 基于纳米材料的电化学传感器在重金属离子检测中的应用[J]. 南京工业大学学报(自然科学版), 2018, 40(3): 115-121.

[21] 杨帅, 汤婉鑫, 张超, 等. PdNPs/MWNTs修饰玻碳电极的制备及其对六价铬的电化学测定[J]. 上海师范大学学报(自然科学版), 2014, 43(6): 594-599.

[22] 朱伟明, 梁新义, 庞广昌, 等. 基于金钯纳米合金修饰的过氧化氢传感器的研制[J]. 食品科学, 2012, 33(10): 311-314.

[23] KUMAR S S, KWAK K, LEE D. Electrochemical sensing using quantum-sized gold nanoparticles[J]. Analytical chemistry, 2011, 83(9): 3244-3247.

[24] ABOLLINO O, GIACOMINO A, MALANDRINO M, et al. Determination of mercury by anodic stripping voltammetry with a gold nanoparticle-modified glassy carbon electrode[J]. Electroanalysis, 2008, 20(1): 75-83.

[25] AHMAD A, WEI Y, SYED F, et al. Size dependent catalytic activities of green synthesized gold nanoparticles and electro-catalytic oxidation of catechol

on gold nanoparticles modified electrode[J]. RSC Advances, 2015, 5(120): 99364-99377.

[26] SANG S B, LI D, ZHANG H, et al. Facile synthesis of AgNPs on reduced graphene oxide for highly sensitive simultaneous detection of heavy metal ions[J]. RSC Advances, 2017, 7(35): 21618-21624.

[27] WU T X, LI T T, LIU Z G, et al. Electrochemical sensor for sensitive detection of triclosan based on graphene/palladium nanoparticles hybrids[J]. Talanta, 2017, 164: 556-562.

[28] ZHANG Q X, YIN L B. Electrochemical performance of heterostructured Au-Pd bimetallic nanoparticles toward As(III) aqueous media[J]. Electrochemistry communications, 2012, 22: 57-60.

[29] 张磊, 王旭, 李忠平, 等. 钯纳米粒子修饰的玻碳电极对Cu^{2+}的电化学检测[J]. 山西大学学报(自然科学版), 2018, 41(4): 781-786.

[30] ZHANG L, LI Z P, ZHANG Y, et al. High-performance liquid chromatography coupled with mass spectrometry for analysis of ultrasmall palladium nanoparticles[J]. Talanta, 2015, 131: 632-639.

[31] KANG Y J, YE X C, MURRAY C B. Size-and shape-selective synthesis of metal nanocrystals and nanowires using co as a reducing agent[J]. Angewandte chemie international edition, 2010, 49(35): 6156-6159.

第7章
总结与展望

7.1 总结

钯纳米材料作为重要的贵金属纳米材料之一，在催化、光吸收和磁性能等方面体现出特有的物理和化学性质，而这些性质与其的微观结构有着千丝万缕的联系，因此，针对钯纳米材料性能、结构分析方面的研究具有重要的意义。本书首先，围绕小尺寸水溶性钯纳米的制备和表征；然后，采用高效液相色谱法应用于Pd NPs的分离，利用质谱法对分离的各个组分进行测定，得到了有机小分子保护的Pd NPs的更加详细和丰富的化学组成信息；最后，将水溶性Pd NPs修饰电极用于重金属铜离子的检测。纵观全文，主要研究结果如下。

7.1.1 NAC保护的Pd NPs的制备和表征

在冰浴条件下，以NAC作为配体，不同NAC/Pd摩尔比条件下，通过$NaBH_4$还原H_2PdCl_2合成了一系列水溶性Pd NPs（NAC-Pd NPs）。通过UV-vis、XPS、FTIR、TGA和TEM等技术对不同摩尔比合成的NAC-Pd NPs进行表征，得出NAC/Pd摩尔比大于2时，所合成的Pd NPs为Pd^{II}-NAC的复合物；当NAC/Pd摩尔比为1∶1合成的NAC-Pd NPs为Pd^0的纳米粒子，粒径为（2.17 ± 0.20）nm，其水溶性、分散性和稳定性好。由此得出，在此反应过程中配体NAC中的巯基与钯原子结合生成稳定的Pd—S键，通过$NaBH_4$还原H_2PdCl_4成为Pd NPs，NAC也迅速与钯原子反应，将其包裹起来，阻止已生成Pd NPs的再聚集，NAC作为配体起到了稳定和保护Pd NPs的作用。

7.1.2 NAC保护的Pd NPs的分离分析

采用色谱柱C18（250 mm × 4.6 mm，5 μm），流动相为甲醇和$Bu_4N^+F^-$水溶液，结合梯度洗脱，利用反向离子对高效液相色谱法分离水溶性NAC-Pd NPs。所分离的11种NAC-Pd NPs组分按照钯原子数目从C18色谱柱洗脱的先后被洗脱出来。研究了$Bu_4N^+F^-$和甲醇含量对NAC-Pd NPs分离的影响；利用UV-vis和MALDI-TOF MS对分离的11种NAC-Pd NPs组分进行分析。研究结果表明，每一个色谱分离组分都具有不同的UV-vis，这些性质通过研究NAC-Pd NPs混合物无法得到的。通过研究NAC-Pd NPs色谱分离组分的MALDI-TOF MS，可显示组分的粒子特性，得到每一个分离组分的化学组成，分别为$Pd_{10}(NAC)_7$、

$Pd_{11}(NAC)_7$、$Pd_{11}(NAC)_8$、$Pd_{12}(NAC)_9$、$Pd_{13}(NAC)_6$、$Pd_{13}(NAC)_9$、$Pd_{14}(NAC)_5$、$Pd_{14}(NAC)_9$、$Pd_{15}(NAC)_9$、$Pd_{17}(NAC)_{11}$和$Pd_{20}(NAC)_{11}$。

7.1.3 DMF保护的Pd NPs的制备和表征

在不同反应时间下,采用一步法合成DMF作为配体保护的水溶性Pd NPs（DMF-Pd NPs）,通过紫外-可见分光光谱、荧光光谱、红外光谱、热重分析和透射电子显微镜等技术对合成的DMF-Pd NPs进行表征。研究结果表明,在合成反应过程中,DMF同时起到三重作用,分别是溶剂、配体和还原剂,DMF在加热温度高于100℃时分解为二甲胺和一氧化碳（CO）,其中CO作为还原剂,将钯离子还原为零价钯原子,而DMF作为配体和钯原子配位形成配合物,包裹在钯纳米外围,防止Pd NPs的聚集,从而形成DMF-Pd NPs。采用此方法合成的DMF-Pd NPs具有荧光发光特性,分散性好,稳定性好,平均粒径为（2.20 ± 0.70）nm,且易溶于水和甲醇等有机试剂。

7.1.4 DMF保护的Pd NPs的分离分析

采用C18色谱柱（250 mm × 4.6 mm,5 μm）,流动相为甲醇和水,结合梯度洗脱,利用反向高效液相色谱法分离水溶性DMF-Pd NPs。研究了流动相中不同甲醇含量对DMF-Pd NPs分离的影响,利用UV-vis、FL和MALDI-TOF MS对所分离的13种DMF-Pd NPs组分进行分析。研究结果表明,每一个分离组分都有自己特征的UV-vis和FL,这些性质与研究DMF-Pd NPs混合物时无法获知的。采用MALDI-TOF-MS对DMF-Pd NPs分离组分进行分析,可显示每一组分的粒子特性,得到分离组分的化学组成,分别为$Pd_{14}(DMF)_8$、$Pd_{10}(DMF)_8$、$Pd_{12}(DMF)_{11}$、$Pd_{14}(DMF)_8$、$Pd_{14}(DMF)_{10}$、$Pd_{14}(DMF)_{12}$、$Pd_{15}(DMF)_{12}$、$Pd_{16}(DMF)_{15}$、$Pd_{16}(DMF)_{15}$、$Pd_{16}(DMF)_{15}$、$Pd_{20}(DMF)_9$、$Pd_{17}(DMF)_{13}$和$Pd_{20}(DMF)_9$。其中组分8～10具有相同的分子式$Pd_{16}(DMF)_{15}$,组分11、组分13既具有相同的分子式$Pd_{20}(DMF)_9$,但是从C18色谱柱洗脱的先后顺序不同,归因于DMF配体与Pd NPs的配位差异。

7.1.5 DMF保护的Pd NPs修饰的玻碳电极对Cu^{2+}的电化学检测

采用滴涂法制备DMF-Pd NPs修饰玻碳电极,并在对修饰电极电化学表征的基础上,通过优化DMF-Pd NPs/GC修饰电极对Cu^{2+}的电化学响应性能,建立了一种基于DMF-Pd NPs/GC修饰电极对Cu^{2+}的检测方法，从而实现了对Cu^{2+}的稳定、快速和灵敏的测定，该电化学传感器在浓度（4×10^{-7}）~（5×10^{-5}）mol/L内铜离子表现出较好的线性关系，检出限为5×10^{-7} mol/L（S/N=3），并且DMF-Pd NPs/GC修饰电极对Cd^{2+}、Zn^{2+}、Mn^{2+}、Co^{2+}、Mg^{2+}、Fe^{2+}、Hg^{2+}、Pb^{2+}等常见的重金属污染物离子几乎没有电化学响应行为，而只是对Cu^{2+}表现出较好的响应性能，说明所制备的DMF-Pd NPs/GC修饰电极对金属Cu^{2+}具有良好的选择性。

7.2 展望

Pd NPs的应用与环境分析化学方面仍然是一个有待深入的课题，本书使用不同的有机配体所做的尝试，为实际运用提供了积极的帮助。但是，在将来的研究工作中还需继续努力，以下方面的研究内容是今后科研工作的重点。

7.2.1 色谱峰分离组分的研究与应用

在本书的研究中，针对有机配体NAC和DMF保护的Pd NPs通过RP-HPLC法分别分离出11个和13个色谱分离峰，并采用UV-vis和MALDI-TOF-MS等技术对所分离的色谱分离峰的化学组成信息进行了详细的分析，得到了Pd NPs分离组分中所包含的钯原子和配体数目。但是，针对所分离的Pd NPs色谱峰组分与未经分离的NAC和DMF保护的Pd NPs催化等的作用是否一致，或者其催化等性能会更加卓越，有待我们进一步探索研究。

7.2.2 DMF-Pd NPs/GC修饰电极对Cu^{2+}的检测机理

在本书的研究中，通过制备DMF-Pd NPs修饰玻碳电极，并在对修饰电极电化学表征的基础上，通过优化DMF-Pd NPs/GC修饰电极对Cu^{2+}的电化学响应性能，建立了一种基于DMF-Pd NPs/GC修饰电极对Cu^{2+}的检测方法，从而实现了对Cu^{2+}的稳定、快速和灵敏的测定。但是，针对DMF-Pd NPs/GC修饰电极对Cu^{2+}的检测的机理只是推测，需要我们在今后的研究工作中，对DMF-Pd NPs/GC修饰电极的检测机理进行更进一步的研究，得到其测定Cu^{2+}的作用机理。

彩 图

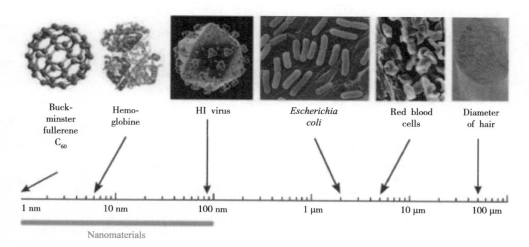

彩图1.1 纳米粒子长度尺寸的分类

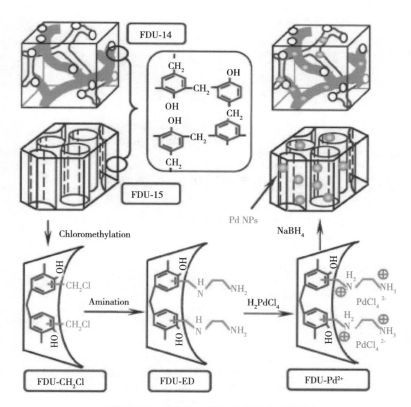

彩图1.6 介孔聚合物合成Pd NPs示意图

有机配体保护的钯纳米粒子的制备及其分析应用

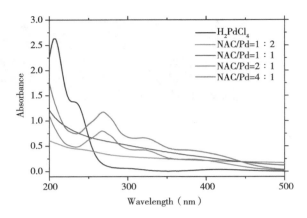

彩图2.2　不同摩尔比合成的NAC-Pd NPs和H_2PdCl_4的UV-vis图

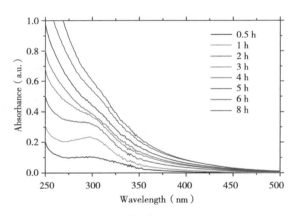

彩图4.2　不同反应时间的DMF-Pd NPs的UV-vis

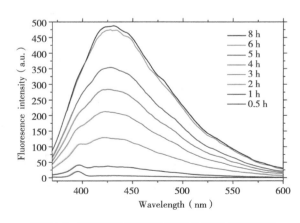

彩图4.4　不同反应时间的DMF-Pd NPs溶液的FL